Martini's Atlas of the Human Body

by

Frederic H. Martini, Ph.D.
University of Hawaii

with

William C. Ober, M.D.
Art Coordinator and Illustrator

Claire W. Garrison, R.N.
Illustrator

Kathleen Welch
Clinical Consultant

Ralph T. Hutchings, Ph.D.
Biomedical Photographer

PEARSON

Benjamin
Cummings

San Francisco   Boston   New York
Cape Town   Hong Kong   London   Madrid   Mexico City
Montreal   Munich   Paris   Singapore   Sidney   Tokyo   Toronto

Editorial Director: *Daryl Fox*
Executive Editor: *Leslie Berriman*
Project Editor: *Nicole George-O'Brien*
Editorial Assistant: *Blythe Robbins*
Project Management: *Elm Street Publishing Services, Inc.*
Compositor: *Preparé, Inc.*
Managing Editor: *Deborah Cogan*
Senior Production Supervisor: *Corinne Benson*
Manufacturing Buyer: *Stacey Weinberger*
Executive Marketing Manager: *Lauren Harp*
Text Printer: *R. R. Donnelley, Willard*
Cover Printer: *Phoenix Color Corporation*
Design Manager: *Mark Ong*
Cover Designer: *Yvo Riezebos*

Library of Congress Cataloging-in-Publishing Data

Martini, Frederic.
  Martini's atlas of the human body / by Frederic H. Martini, with William C. Ober ... [et al.].—7th ed.
    p. ; cm.
  ISBN 0-8053-7287-3 (alk. paper)
  1. Human anatomy—Atlases.
  [DNLM: 1. Anatomy—Atlases. 2. Physiology—Atlases. QS 17 M386m 2005] I. Title: Atlas of the human body. II. Ober,
William C. III. Fundamentals of anatomy and physiology. IV. Title.
  QM25.M286 2006
  611—dc22   2004024369

ISBN: 0-8053-7287-3

PEARSON

Benjamin
Cummings

www.aw-bc.com   3 4 5 6 7 8 9 10—DOW—07 06 05

Contents

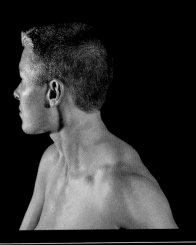 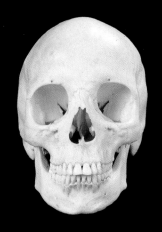 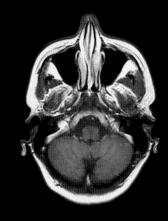

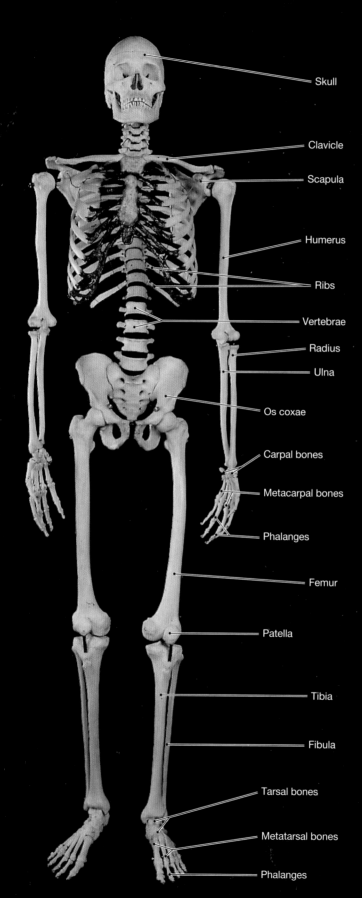

Skull

Clavicle

Scapula

Humerus

Ribs

Vertebrae

Radius

Ulna

Os coxae

Carpal bones

Metacarpal bones

Phalanges

Femur

Patella

Tibia

Fibula

Tarsal bones

Metatarsal bones

Phalanges

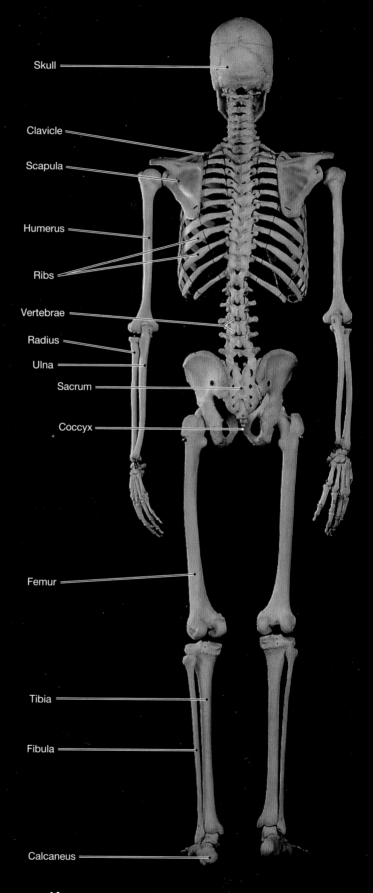

Skull

Clavicle

Scapula

Humerus

Ribs

Vertebrae

Radius

Ulna

Sacrum

Coccyx

Femur

Tibia

Fibula

Calcaneus

PLATE **1a** THE SKELETON, ANTERIOR VIEW

PLATE **1b** THE SKELETON, POSTERIOR VIEW

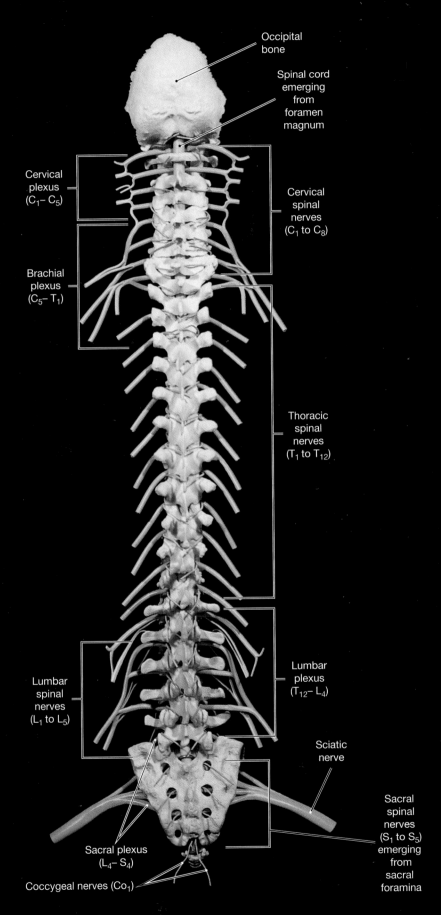

Occipital
bone

Spinal cord
emerging
from
foramen
magnum

Cervical
plexus
(C₁– C₅)

Cervical
spinal
nerves
(C₁ to C₈)

Brachial
plexus
(C₅– T₁)

Thoracic
spinal
nerves
(T₁ to T₁₂)

Lumbar
spinal
nerves
(L₁ to L₅)

Lumbar
plexus
(T₁₂– L₄)

Sciatic
nerve

Sacral
spinal
nerves
(S₁ to S₅)
emerging
from
sacral
foramina

Sacral plexus
(L₄– S₄)

Coccygeal nerves (Co₁)

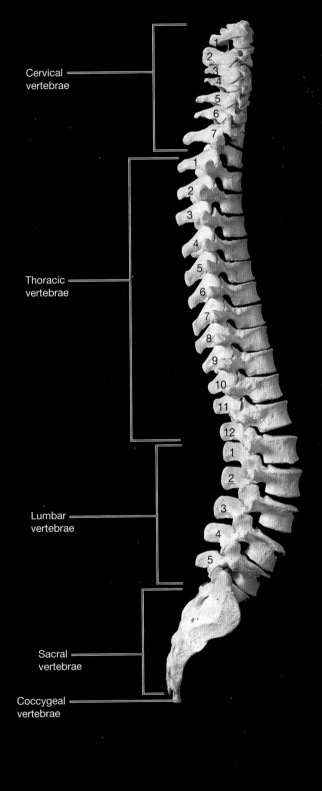

Cervical
vertebrae

Thoracic
vertebrae

Lumbar
vertebrae

Sacral
vertebrae

Coccygeal
vertebrae

PLATE 2a THE VERTEBRAL COLUMN AND SPINAL NERVES   **PLATE 2b** THE VERTEBRAL COLUMN, LATERAL VIEW

PLATE **3a** SURFACE ANATOMY OF THE
HEAD AND NECK, ANTERIOR VIEW

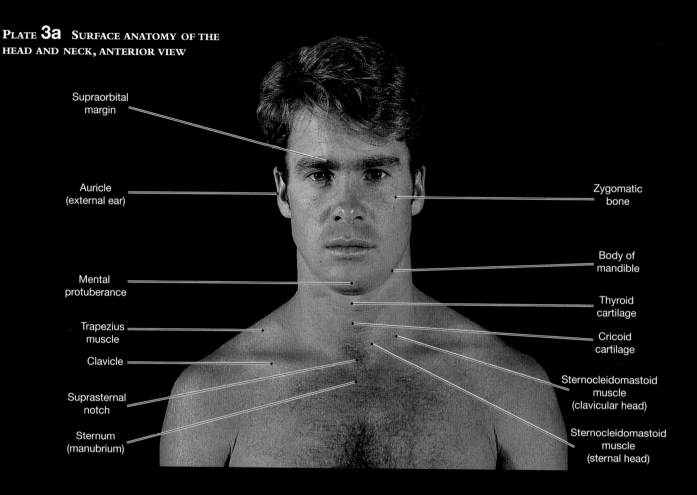

Supraorbital margin

Auricle (external ear)

Zygomatic bone

Body of mandible

Mental protuberance

Thyroid cartilage

Trapezius muscle

Cricoid cartilage

Clavicle

Sternocleidomastoid muscle (clavicular head)

Suprasternal notch

Sternum (manubrium)

Sternocleidomastoid muscle (sternal head)

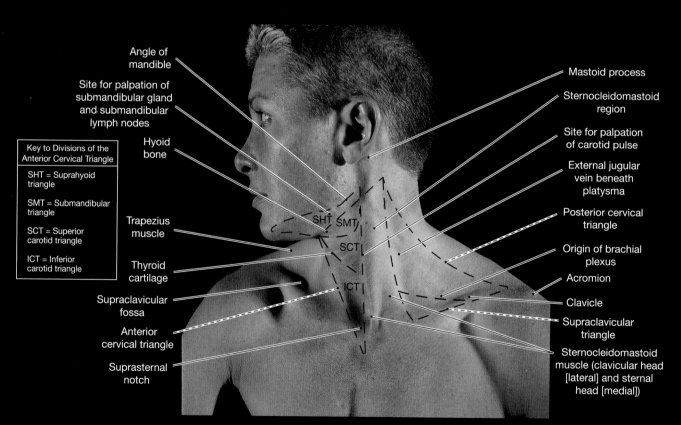

Angle of mandible

Site for palpation of submandibular gland and submandibular lymph nodes

Mastoid process

Sternocleidomastoid region

Site for palpation of carotid pulse

External jugular vein beneath platysma

Hyoid bone

Key to Divisions of the Anterior Cervical Triangle

SHT = Suprahyoid triangle

SMT = Submandibular triangle

SCT = Superior carotid triangle

ICT = Inferior carotid triangle

Trapezius muscle

Posterior cervical triangle

Origin of brachial plexus

Thyroid cartilage

Acromion

Supraclavicular fossa

Clavicle

Anterior cervical triangle

Supraclavicular triangle

Suprasternal notch

Sternocleidomastoid muscle (clavicular head [lateral] and sternal head [medial])

SHT SMT
SCT
ICT

PLATE **3b** THE CERVICAL TRIANGLES

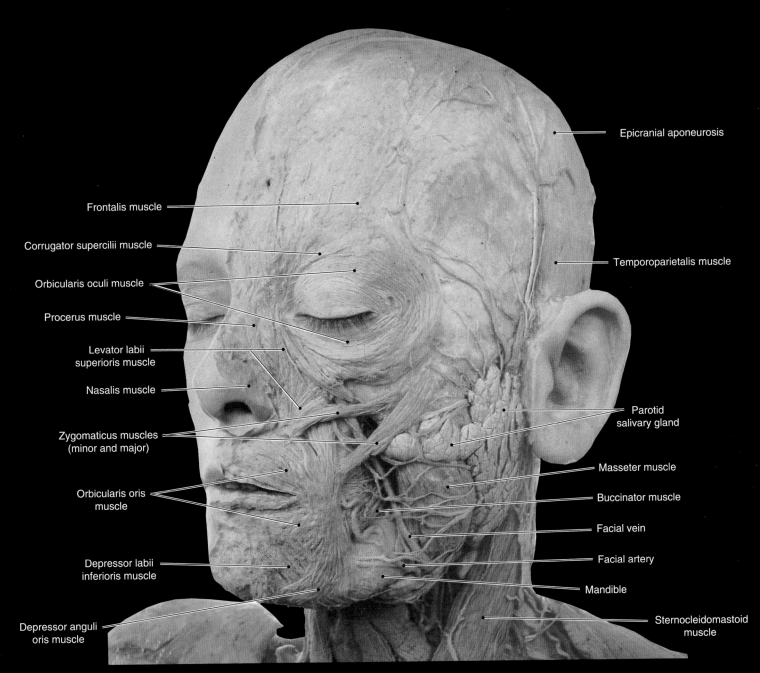

Epicranial aponeurosis

Frontalis muscle

Corrugator supercilii muscle

Orbicularis oculi muscle

Procerus muscle

Levator labii
superioris muscle

Nasalis muscle

Zygomaticus muscles
(minor and major)

Orbicularis oris
muscle

Depressor labii
inferioris muscle

Depressor anguli
oris muscle

Temporoparietalis muscle

Parotid
salivary gland

Masseter muscle

Buccinator muscle

Facial vein

Facial artery

Mandible

Sternocleidomastoid
muscle

PLATE 3C SUPERFICIAL DISSECTION OF THE FACE, ANTEROLATERAL VIEW

PLATE **3d** SUPERFICIAL DISSECTION OF THE FACE, LATERAL VIEW

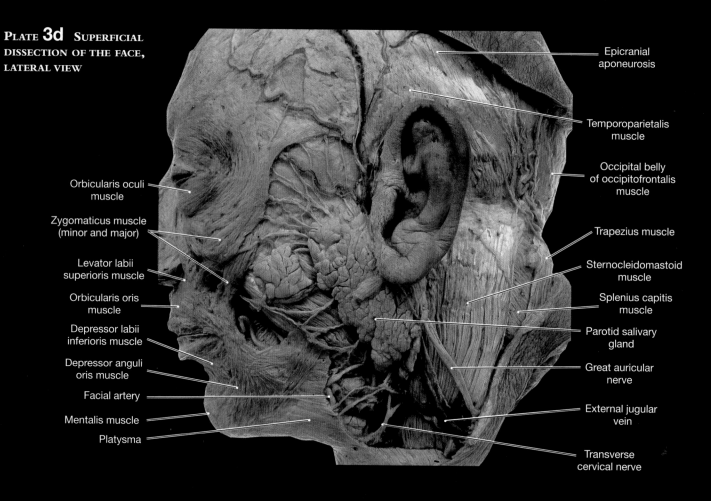

Epicranial aponeurosis

Temporoparietalis muscle

Occipital belly of occipitofrontalis muscle

Orbicularis oculi muscle

Zygomaticus muscle (minor and major)

Levator labii superioris muscle

Orbicularis oris muscle

Depressor labii inferioris muscle

Depressor anguli oris muscle

Facial artery

Mentalis muscle

Platysma

Trapezius muscle

Sternocleidomastoid muscle

Splenius capitis muscle

Parotid salivary gland

Great auricular nerve

External jugular vein

Transverse cervical nerve

PLATE **4a** PAINTED SKULL, LATERAL VIEW

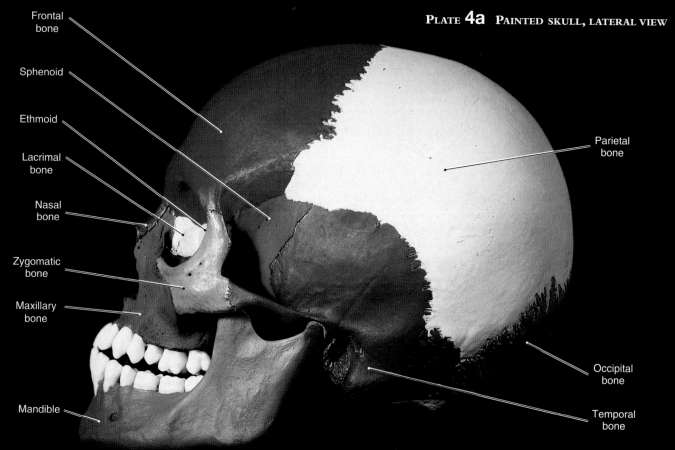

Frontal bone

Sphenoid

Ethmoid

Lacrimal bone

Nasal bone

Zygomatic bone

Maxillary bone

Mandible

Parietal bone

Occipital bone

Temporal bone

PLATE **4b** PAINTED SKULL,
ANTEROLATERAL VIEW

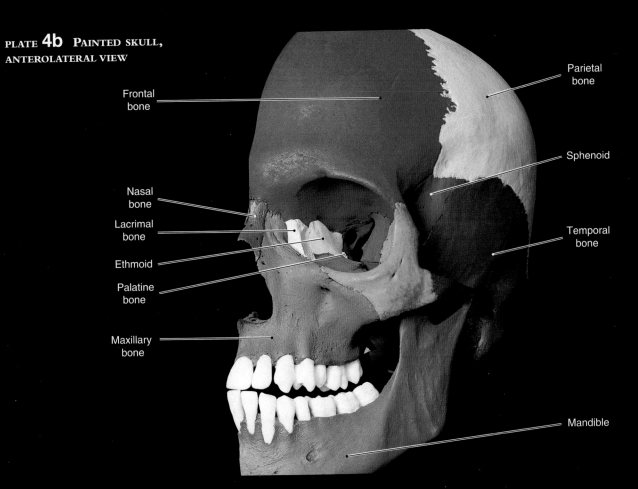

Frontal
bone

Parietal
bone

Sphenoid

Nasal
bone

Lacrimal
bone

Ethmoid

Palatine
bone

Temporal
bone

Maxillary
bone

Mandible

PLATE **4c** PAINTED SKULL,
MEDIAL VIEW

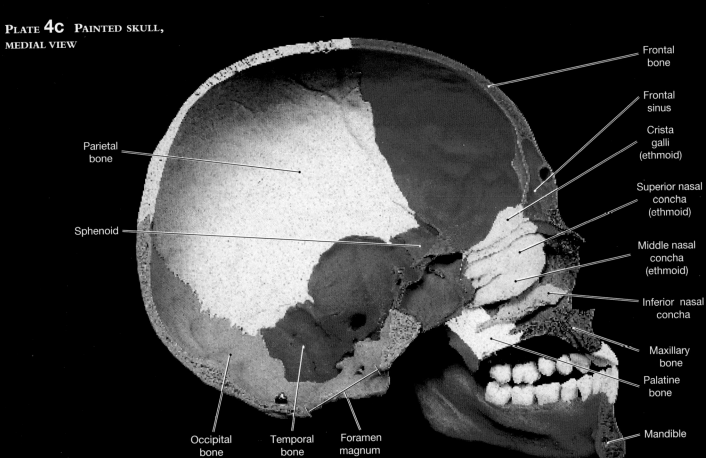

Frontal
bone

Frontal
sinus

Crista
galli
(ethmoid)

Superior nasal
concha
(ethmoid)

Parietal
bone

Middle nasal
concha
(ethmoid)

Sphenoid

Inferior nasal
concha

Maxillary
bone

Palatine
bone

Occipital
bone

Temporal
bone

Foramen
magnum

Mandible

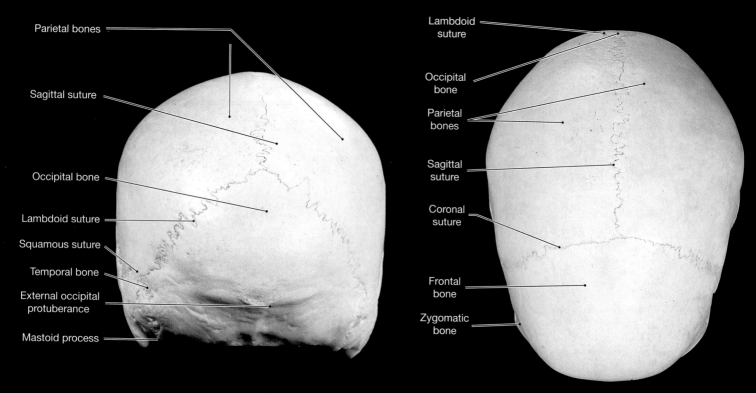

Parietal bones

Sagittal suture

Occipital bone

Lambdoid suture

Squamous suture

Temporal bone

External occipital protuberance

Mastoid process

Lambdoid suture

Occipital bone

Parietal bones

Sagittal suture

Coronal suture

Frontal bone

Zygomatic bone

PLATE **5a** ADULT SKULL, POSTERIOR VIEW

PLATE **5b** ADULT SKULL, SUPERIOR VIEW

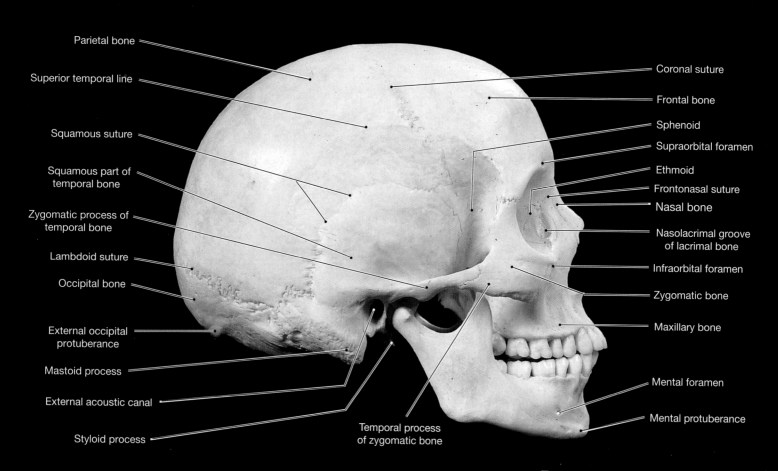

Parietal bone

Superior temporal line

Squamous suture

Squamous part of temporal bone

Zygomatic process of temporal bone

Lambdoid suture

Occipital bone

External occipital protuberance

Mastoid process

External acoustic canal

Styloid process

Coronal suture

Frontal bone

Sphenoid

Supraorbital foramen

Ethmoid

Frontonasal suture

Nasal bone

Nasolacrimal groove of lacrimal bone

Infraorbital foramen

Zygomatic bone

Maxillary bone

Mental foramen

Mental protuberance

Temporal process of zygomatic bone

PLATE **5c** ADULT SKULL, LATERAL VIEW

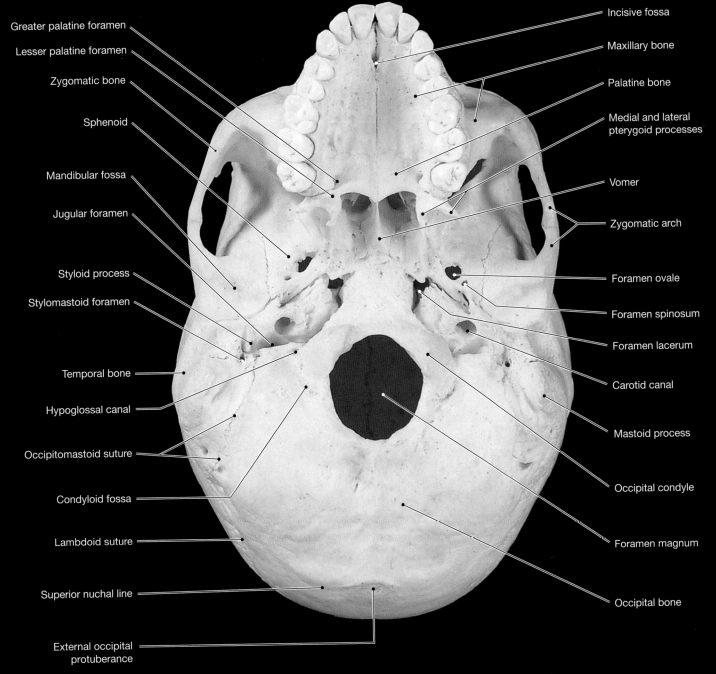

Incisive fossa

Greater palatine foramen

Lesser palatine foramen

Zygomatic bone

Sphenoid

Mandibular fossa

Jugular foramen

Styloid process

Stylomastoid foramen

Temporal bone

Hypoglossal canal

Occipitomastoid suture

Condyloid fossa

Lambdoid suture

Superior nuchal line

External occipital
protuberance

Maxillary bone

Palatine bone

Medial and lateral
pterygoid processes

Vomer

Zygomatic arch

Foramen ovale

Foramen spinosum

Foramen lacerum

Carotid canal

Mastoid process

Occipital condyle

Foramen magnum

Occipital bone

PLATE 5d ADULT SKULL, INFERIOR VIEW (MANDIBLE REMOVED)

PLATE **5e**   ADULT SKULL, ANTERIOR VIEW

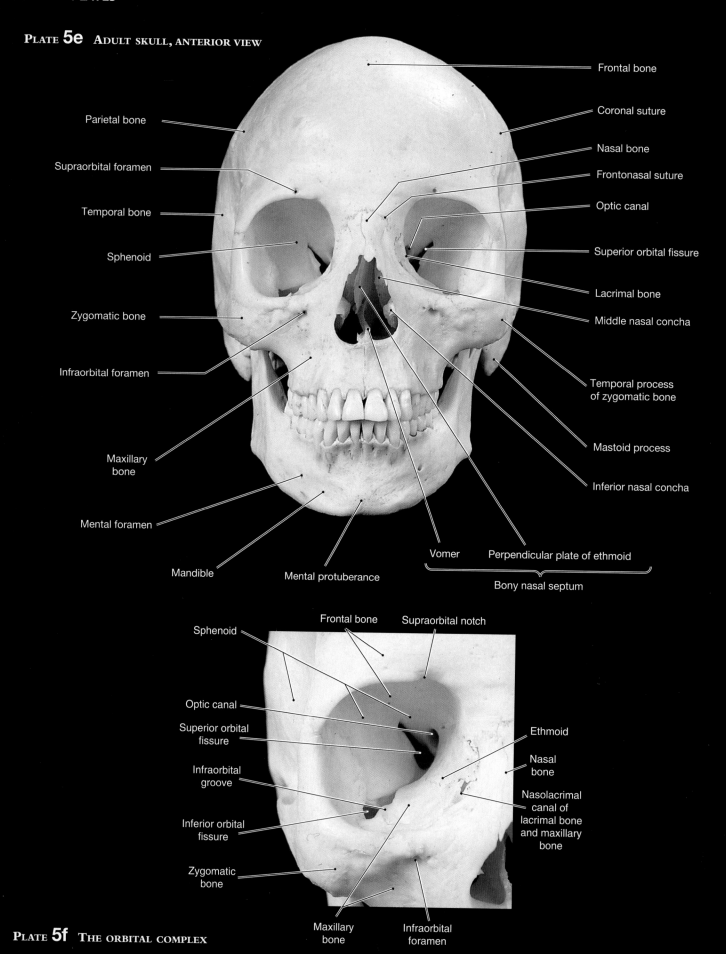

Frontal bone

Parietal bone

Coronal suture

Supraorbital foramen

Nasal bone

Temporal bone

Frontonasal suture

Sphenoid

Optic canal

Superior orbital fissure

Zygomatic bone

Lacrimal bone

Middle nasal concha

Infraorbital foramen

Temporal process
of zygomatic bone

Mastoid process

Maxillary
bone

Inferior nasal concha

Mental foramen

Mandible   Mental protuberance   Vomer   Perpendicular plate of ethmoid

Bony nasal septum

Sphenoid   Frontal bone   Supraorbital notch

Optic canal

Superior orbital
fissure

Ethmoid

Infraorbital
groove

Nasal
bone

Inferior orbital
fissure

Nasolacrimal
canal of
lacrimal bone
and maxillary
bone

Zygomatic
bone

Maxillary
bone

Infraorbital
foramen

PLATE **5f**   THE ORBITAL COMPLEX

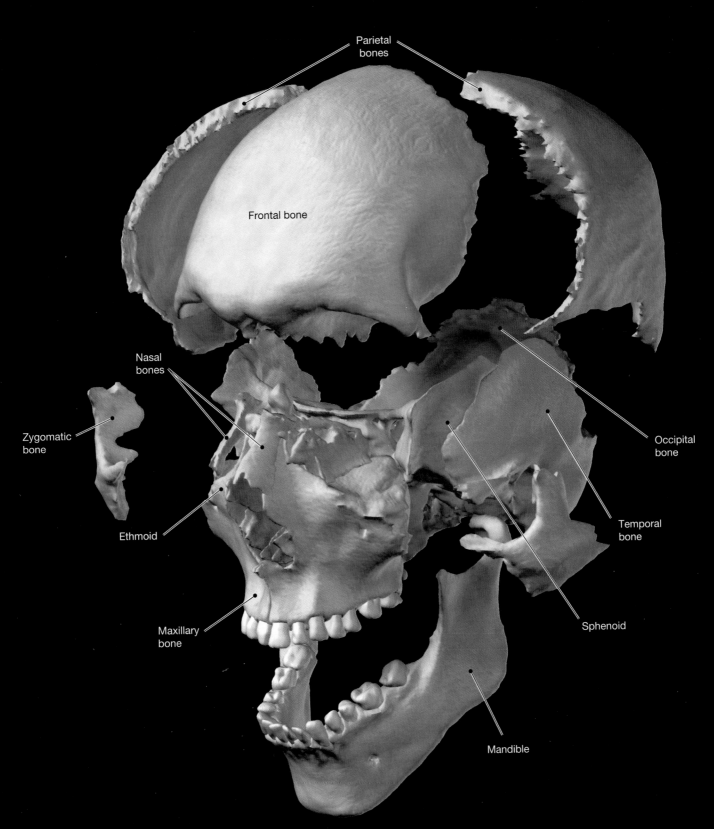

Parietal
bones

Frontal bone

Nasal
bones

Zygomatic
bone

Ethmoid

Maxillary
bone

Occipital
bone

Temporal
bone

Sphenoid

Mandible

PLATE **6** A COMPUTER RECONSTRUCTION OF A DISARTICULATED SKULL

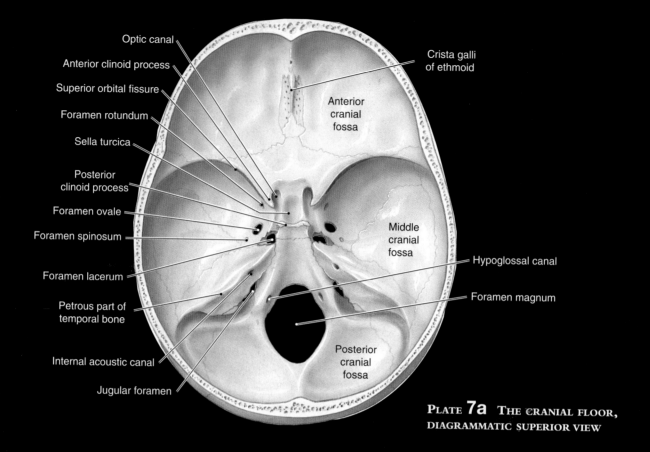

Optic canal

Anterior clinoid process

Superior orbital fissure

Foramen rotundum

Sella turcica

Posterior
clinoid process

Foramen ovale

Foramen spinosum

Foramen lacerum

Petrous part of
temporal bone

Internal acoustic canal

Jugular foramen

Crista galli
of ethmoid

Anterior
cranial
fossa

Middle
cranial
fossa

Hypoglossal canal

Foramen magnum

Posterior
cranial
fossa

PLATE 7a THE CRANIAL FLOOR,
DIAGRAMMATIC SUPERIOR VIEW

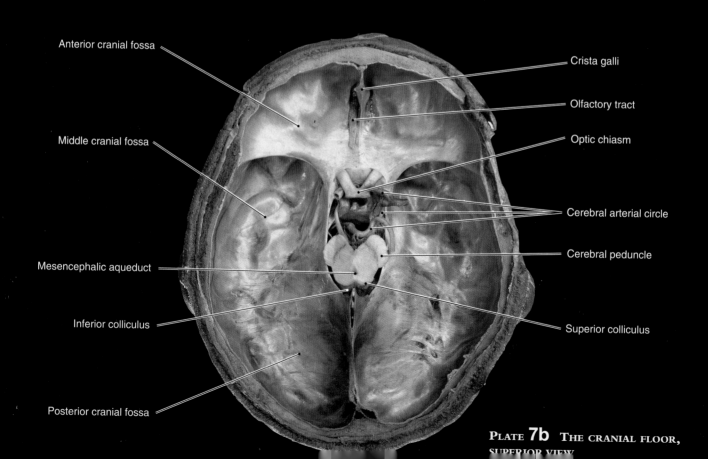

Anterior cranial fossa

Middle cranial fossa

Mesencephalic aqueduct

Inferior colliculus

Posterior cranial fossa

Crista galli

Olfactory tract

Optic chiasm

Cerebral arterial circle

Cerebral peduncle

Superior colliculus

PLATE 7b THE CRANIAL FLOOR,
SUPERIOR VIEW

PLATE 7c THE CRANIAL MENINGES, SUPERIOR VIEW

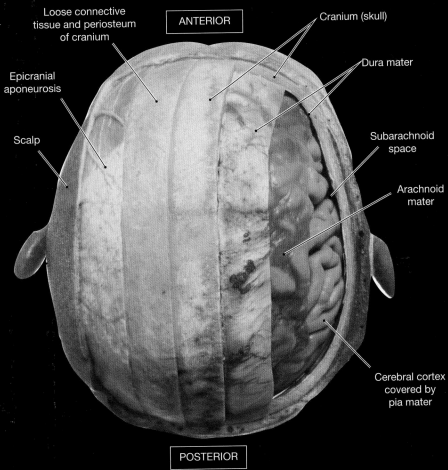

Loose connective tissue and periosteum of cranium

ANTERIOR

Cranium (skull)

Dura mater

Epicranial aponeurosis

Scalp

Subarachnoid space

Arachnoid mater

Cerebral cortex covered by pia mater

POSTERIOR

PLATE 7d DURAL FOLDS

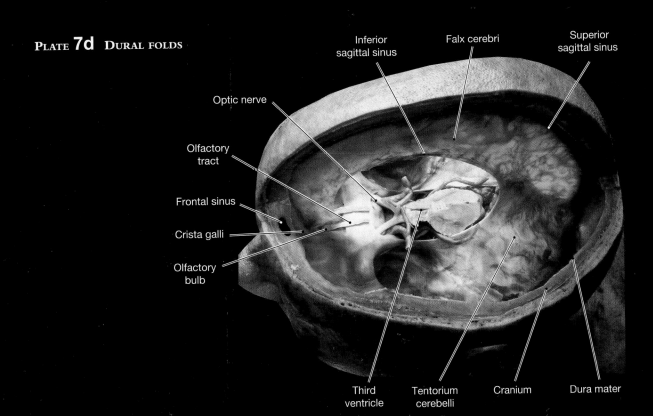

Inferior sagittal sinus

Falx cerebri

Superior sagittal sinus

Optic nerve

Olfactory tract

Frontal sinus

Crista galli

Olfactory bulb

Third ventricle

Tentorium cerebelli

Cranium

Dura mater

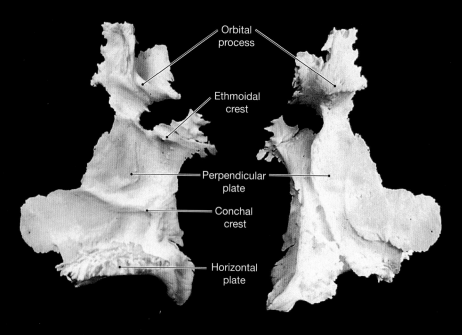

Orbital process

Ethmoidal crest

Perpendicular plate

Conchal crest

Horizontal plate

PLATES **8a–b** PALATINE BONE, MEDIAL AND LATERAL VIEWS

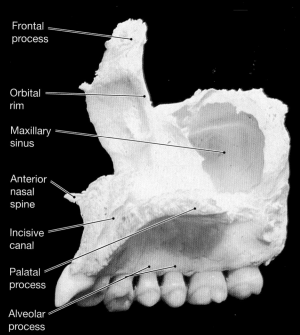

Frontal process

Orbital rim

Maxillary sinus

Anterior nasal spine

Incisive canal

Palatal process

Alveolar process

PLATE **8c** RIGHT MAXILLARY BONE, MEDIAL VIEW

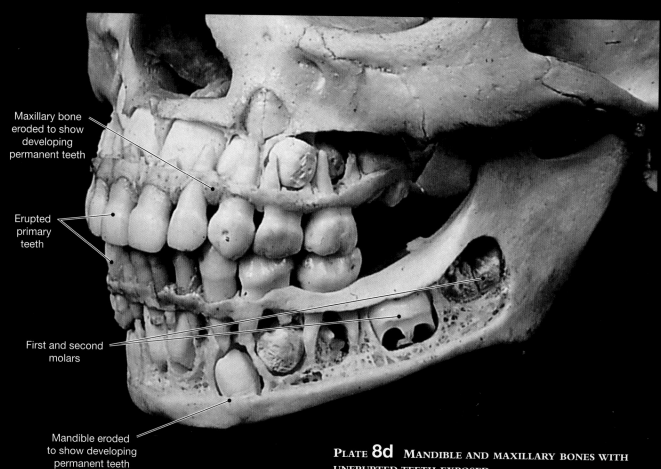

Maxillary bone eroded to show developing permanent teeth

Erupted primary teeth

First and second molars

Mandible eroded to show developing permanent teeth

PLATE **8d** MANDIBLE AND MAXILLARY BONES WITH UNERUPTED TEETH EXPOSED

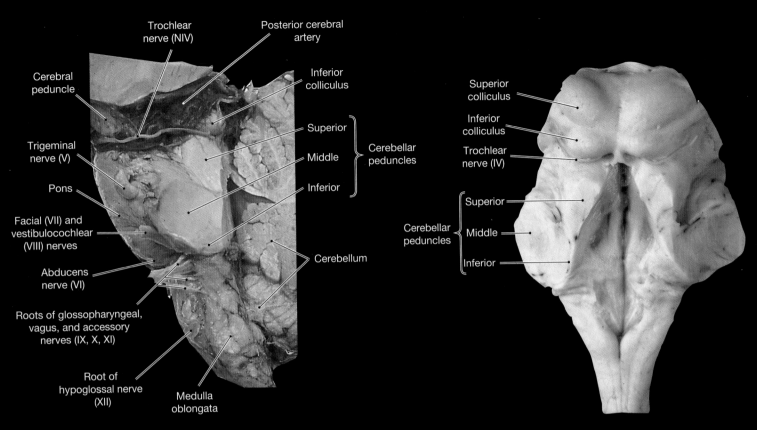

Trochlear nerve (NIV)

Posterior cerebral artery

Cerebral peduncle

Inferior colliculus

Trigeminal nerve (V)

Superior

Middle

Inferior

Cerebellar peduncles

Pons

Facial (VII) and vestibulocochlear (VIII) nerves

Abducens nerve (VI)

Cerebellum

Roots of glossopharyngeal, vagus, and accessory nerves (IX, X, XI)

Root of hypoglossal nerve (XII)

Medulla oblongata

PLATE 9a BRAIN STEM, LATERAL VIEW

Superior colliculus

Inferior colliculus

Trochlear nerve (IV)

Superior

Cerebellar peduncles

Middle

Inferior

PLATE 9b BRAIN STEM, POSTERIOR VIEW

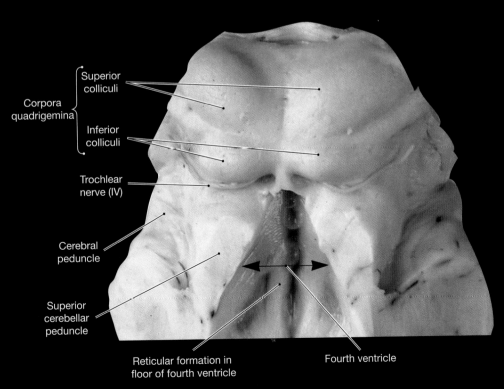

Superior colliculi

Corpora quadrigemina

Inferior colliculi

Trochlear nerve (IV)

Cerebral peduncle

Superior cerebellar peduncle

Reticular formation in floor of fourth ventricle

Fourth ventricle

PLATE 9c THE MESENCEPHALON, POSTERIOR VIEW

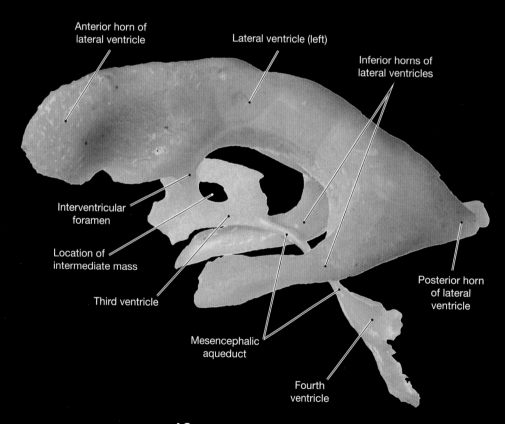

Anterior horn of
lateral ventricle

Lateral ventricle (left)

Inferior horns of
lateral ventricles

Interventricular
foramen

Location of
intermediate mass

Third ventricle

Posterior horn
of lateral
ventricle

Mesencephalic
aqueduct

Fourth
ventricle

PLATE **10** VENTRICLE CASTING, LATERAL VIEW

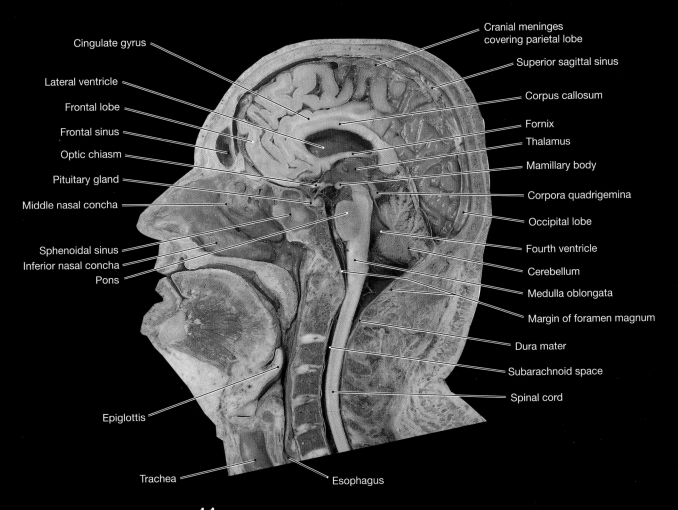

Cingulate gyrus
Lateral ventricle
Frontal lobe
Frontal sinus
Optic chiasm
Pituitary gland
Middle nasal concha
Sphenoidal sinus
Inferior nasal concha
Pons
Epiglottis
Trachea
Esophagus

Cranial meninges
covering parietal lobe
Superior sagittal sinus
Corpus callosum
Fornix
Thalamus
Mamillary body
Corpora quadrigemina
Occipital lobe
Fourth ventricle
Cerebellum
Medulla oblongata
Margin of foramen magnum
Dura mater
Subarachnoid space
Spinal cord

PLATE 11a MIDSAGITTAL SECTION THROUGH THE HEAD AND NECK

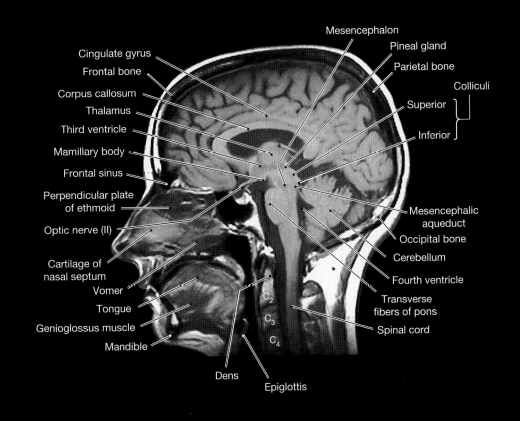

Mesencephalon
Pineal gland
Parietal bone
Colliculi
Superior
Inferior

Cingulate gyrus
Frontal bone
Corpus callosum
Thalamus
Third ventricle
Mamillary body
Frontal sinus
Perpendicular plate
of ethmoid
Optic nerve (II)
Cartilage of
nasal septum
Vomer
Tongue
Genioglossus muscle
Mandible

Mesencephalic
aqueduct
Occipital bone
Cerebellum
Fourth ventricle
Transverse
fibers of pons
Spinal cord

C₂
C₃
C₄

Dens
Epiglottis

**PLATE 11b MRI SCAN OF THE
BRAIN, MIDSAGITTAL SECTION**

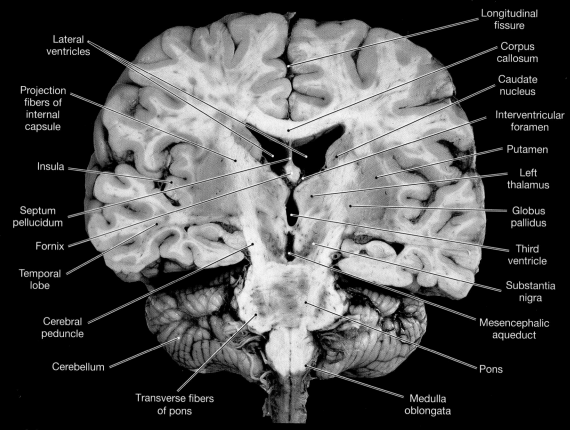

Longitudinal fissure

Corpus callosum

Caudate nucleus

Interventricular foramen

Putamen

Left thalamus

Globus pallidus

Third ventricle

Substantia nigra

Mesencephalic aqueduct

Pons

Medulla oblongata

Lateral ventricles

Projection fibers of internal capsule

Insula

Septum pellucidum

Fornix

Temporal lobe

Cerebral peduncle

Cerebellum

Transverse fibers of pons

PLATE **11c** CORONAL SECTION THROUGH THE BRAIN AT THE LEVEL OF THE MESENCEPHALON AND PONS

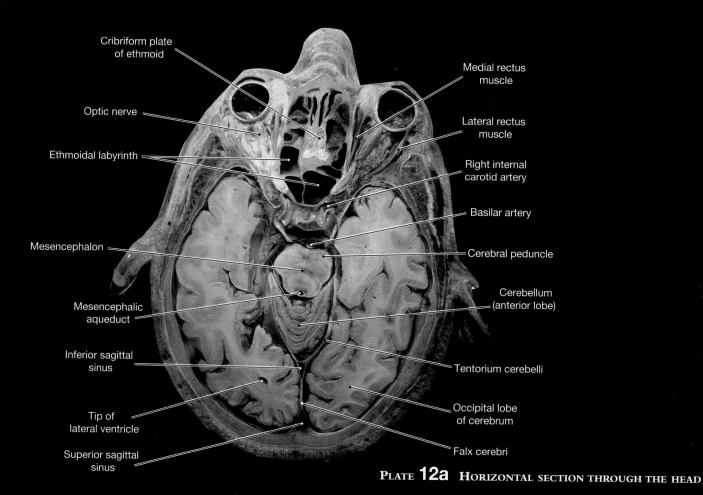

Cribriform plate of ethmoid

Optic nerve

Ethmoidal labyrinth

Mesencephalon

Mesencephalic aqueduct

Inferior sagittal sinus

Tip of lateral ventricle

Superior sagittal sinus

Medial rectus muscle

Lateral rectus muscle

Right internal carotid artery

Basilar artery

Cerebral peduncle

Cerebellum (anterior lobe)

Tentorium cerebelli

Occipital lobe of cerebrum

Falx cerebri

PLATE **12a** HORIZONTAL SECTION THROUGH THE HEAD

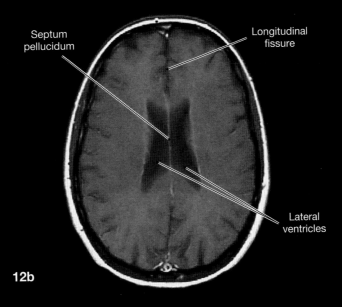

Septum
pellucidum

Longitudinal
fissure

Lateral
ventricles

12b

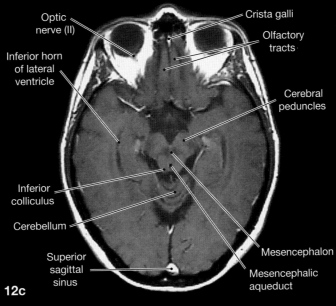

Optic
nerve (II)

Crista galli

Olfactory
tracts

Inferior horn
of lateral
ventricle

Cerebral
peduncles

Inferior
colliculus

Cerebellum

Superior
sagittal
sinus

Mesencephalon

Mesencephalic
aqueduct

12c

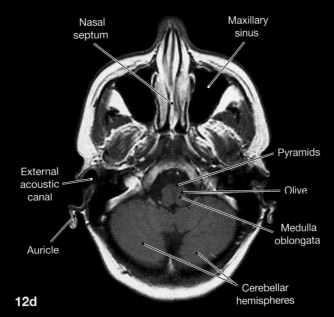

Nasal
septum

Maxillary
sinus

Pyramids

External
acoustic
canal

Olive

Auricle

Medulla
oblongata

Cerebellar
hemispheres

12d

PLATES **12b–d** MRI SCANS
OF THE BRAIN, HORIZONTAL
SECTIONS, SUPERIOR TO
INFERIOR SEQUENCE

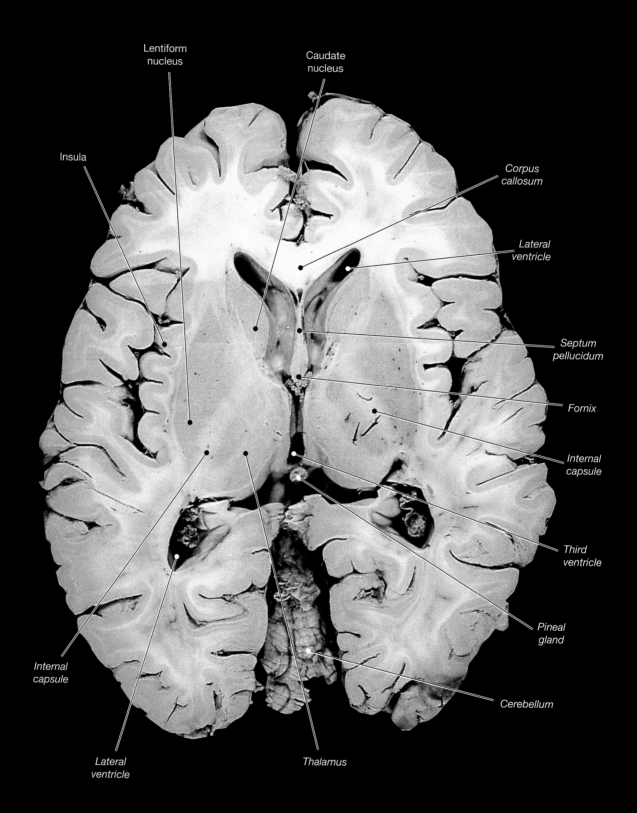

Lentiform nucleus

Caudate nucleus

Insula

Corpus callosum

Lateral ventricle

Septum pellucidum

Fornix

Internal capsule

Third ventricle

Pineal gland

Internal capsule

Cerebellum

Lateral ventricle

Thalamus

PLATE 13a A HORIZONTAL SECTION THROUGH THE BRAIN

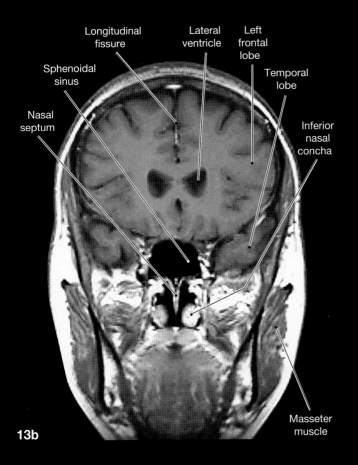

13b

Longitudinal fissure

Lateral ventricle

Left frontal lobe

Temporal lobe

Inferior nasal concha

Sphenoidal sinus

Nasal septum

Masseter muscle

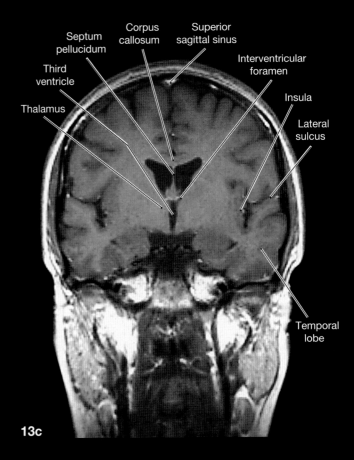

13c

Septum pellucidum

Corpus callosum

Superior sagittal sinus

Interventricular foramen

Insula

Lateral sulcus

Third ventricle

Thalamus

Temporal lobe

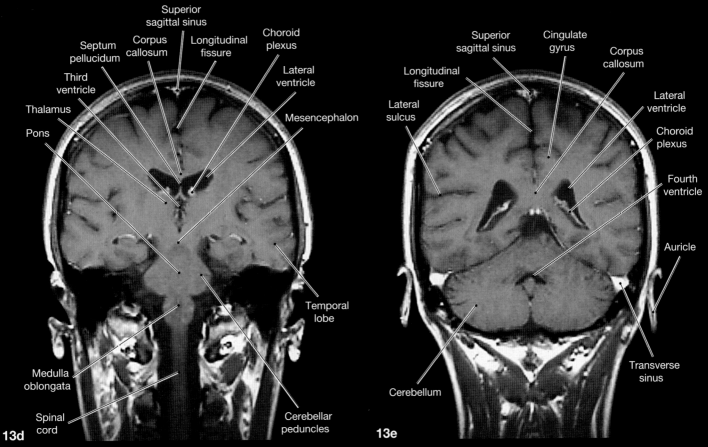

13d

Superior sagittal sinus

Corpus callosum

Longitudinal fissure

Choroid plexus

Lateral ventricle

Mesencephalon

Septum pellucidum

Third ventricle

Thalamus

Pons

Temporal lobe

Medulla oblongata

Spinal cord

Cerebellar peduncles

13e

Superior sagittal sinus

Cingulate gyrus

Corpus callosum

Longitudinal fissure

Lateral ventricle

Choroid plexus

Lateral sulcus

Fourth ventricle

Auricle

Cerebellum

Transverse sinus

PLATES 13b–e MRI SCANS OF THE BRAIN, FRONTAL (CORONAL) SECTIONS, ANTERIOR TO POSTERIOR SEQUENCE

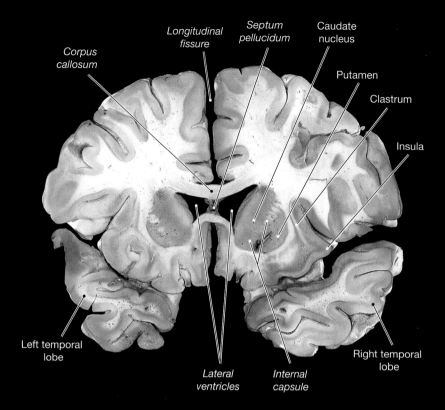

Corpus callosum

Longitudinal fissure

Septum pellucidum

Caudate nucleus

Putamen

Clastrum

Insula

Left temporal lobe

Lateral ventricles

Internal capsule

Right temporal lobe

PLATE 13f FRONTAL SECTION THROUGH THE BRAIN AT THE LEVEL OF THE BASAL NUCLEI

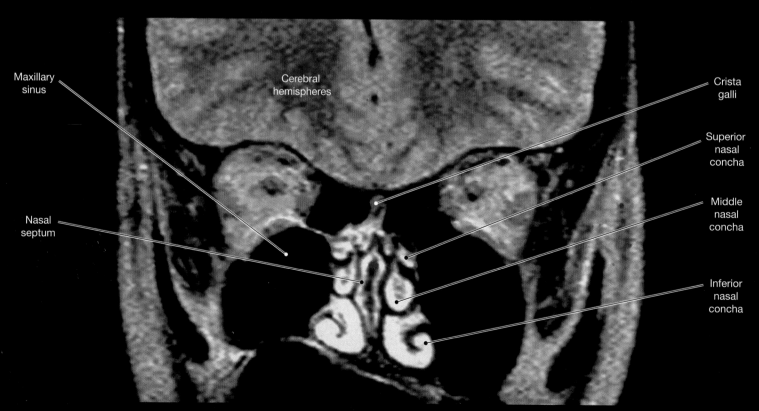

Maxillary sinus

Cerebral hemispheres

Crista galli

Superior nasal concha

Middle nasal concha

Nasal septum

Inferior nasal concha

PLATE 13g MRI SCAN, CORONAL SECTION SHOWING PARANASAL SINUSES

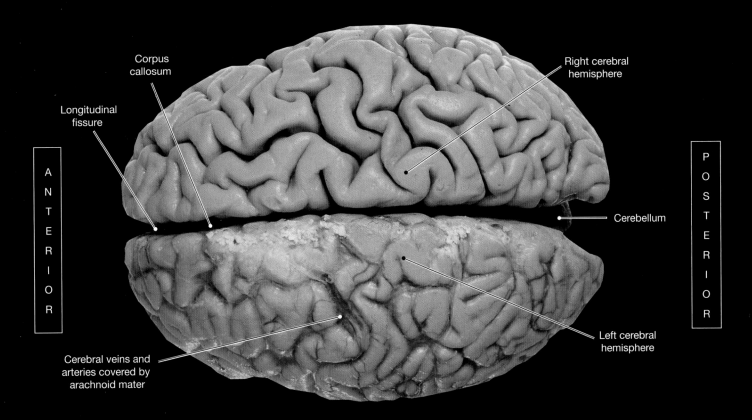

Corpus callosum

Longitudinal fissure

Right cerebral hemisphere

A N T E R I O R

P O S T E R I O R

Cerebellum

Cerebral veins and arteries covered by arachnoid mater

Left cerebral hemisphere

PLATE **14a** SURFACE ANATOMY OF THE BRAIN, SUPERIOR VIEW

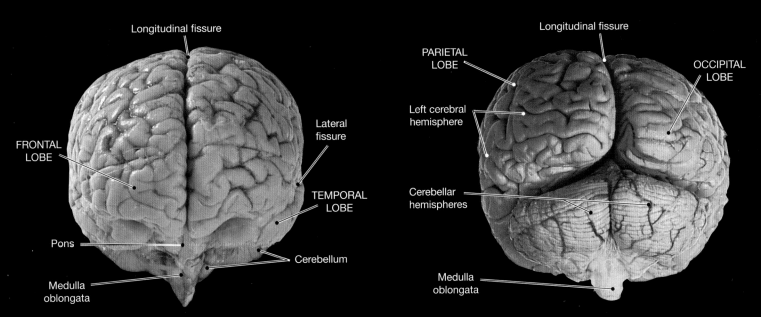

Longitudinal fissure

FRONTAL LOBE

Lateral fissure

TEMPORAL LOBE

Pons

Cerebellum

Medulla oblongata

PLATE **14b** SURFACE ANATOMY OF THE BRAIN, ANTERIOR VIEW

Longitudinal fissure

PARIETAL LOBE

OCCIPITAL LOBE

Left cerebral hemisphere

Cerebellar hemispheres

Medulla oblongata

PLATE **14c** SURFACE ANATOMY OF THE BRAIN, POSTERIOR VIEW

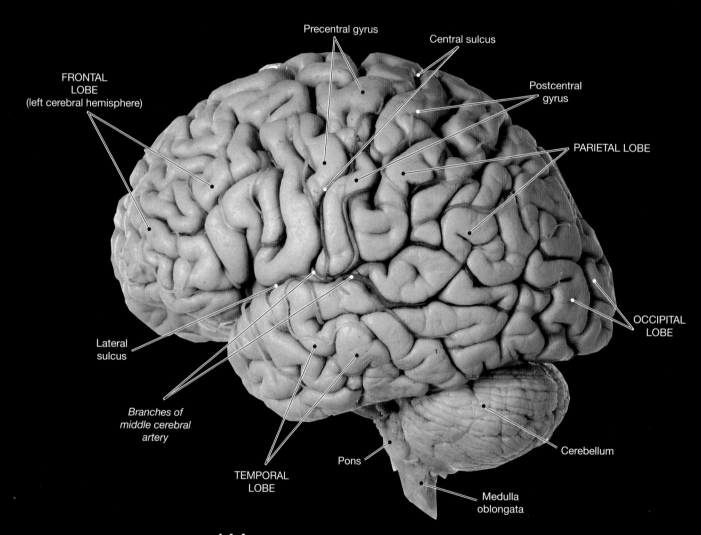

Precentral gyrus

Central sulcus

FRONTAL
LOBE
(left cerebral hemisphere)

Postcentral
gyrus

PARIETAL LOBE

OCCIPITAL
LOBE

Lateral
sulcus

*Branches of
middle cerebral
artery*

Cerebellum

Pons

TEMPORAL
LOBE

Medulla
oblongata

PLATE 14d   SURFACE ANATOMY OF THE BRAIN, LATERAL VIEW

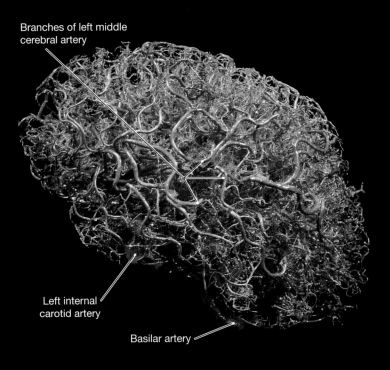

Branches of left middle cerebral artery

Left internal carotid artery

Basilar artery

PLATE 15a ARTERIAL CIRCULATION TO THE BRAIN, LATERAL VIEW OF CORROSION CAST

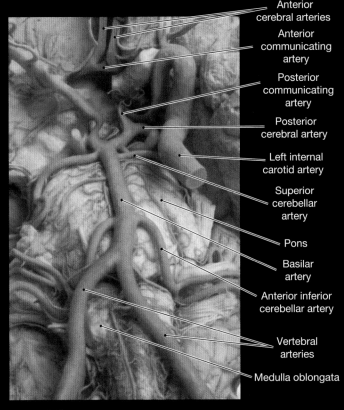

Anterior cerebral arteries

Anterior communicating artery

Posterior communicating artery

Posterior cerebral artery

Left internal carotid artery

Superior cerebellar artery

Pons

Basilar artery

Anterior inferior cerebellar artery

Vertebral arteries

Medulla oblongata

PLATE 15c ARTERIES ON THE INFERIOR SURFACE OF THE BRAIN

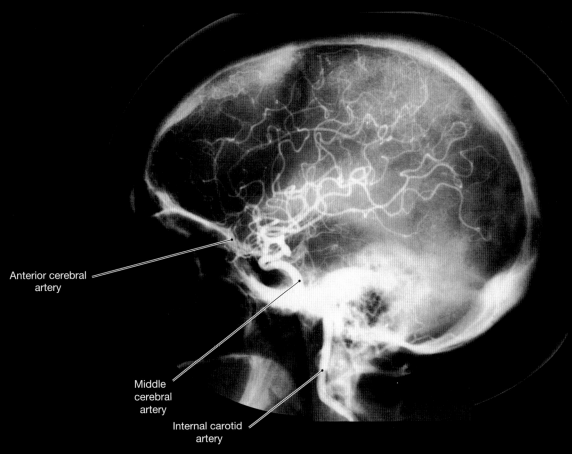

Anterior cerebral artery

Middle cerebral artery

Internal carotid artery

PLATE 15b CRANIAL ANGIOGRAM

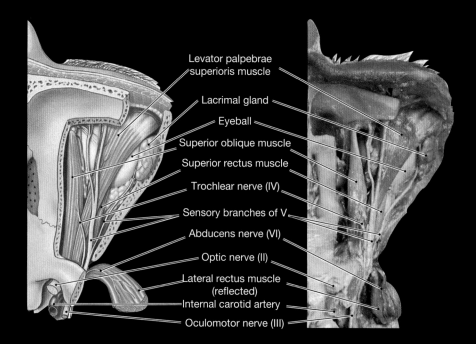

Levator palpebrae superioris muscle

Lacrimal gland

Eyeball

Superior oblique muscle

Superior rectus muscle

Trochlear nerve (IV)

Sensory branches of V

Abducens nerve (VI)

Optic nerve (II)

Lateral rectus muscle (reflected)

Internal carotid artery

Oculomotor nerve (III)

PLATES **16a–b** ACCESSORY STRUCTURES OF THE EYE, SUPERIOR VIEW

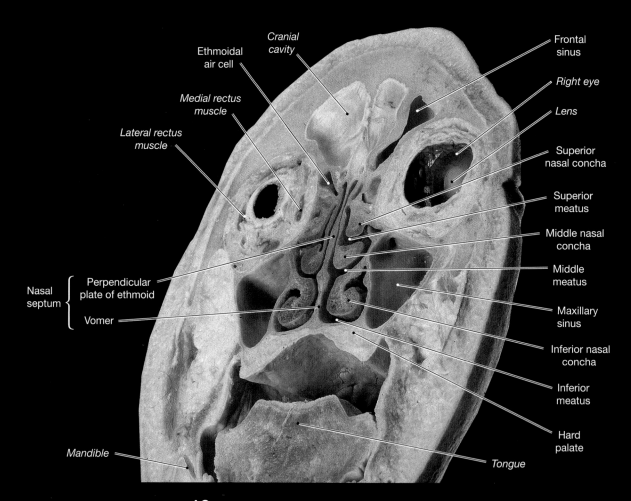

Ethmoidal air cell

Cranial cavity

Frontal sinus

Medial rectus muscle

Right eye

Lens

Lateral rectus muscle

Superior nasal concha

Superior meatus

Middle nasal concha

Middle meatus

Nasal septum { Perpendicular plate of ethmoid

Vomer

Maxillary sinus

Inferior nasal concha

Inferior meatus

Hard palate

Mandible

Tongue

PLATE **16c** FRONTAL SECTION THROUGH THE FACE

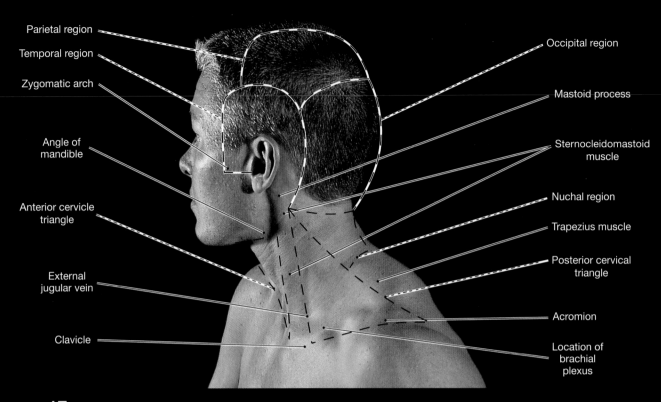

Parietal region

Temporal region

Zygomatic arch

Angle of mandible

Anterior cervicle triangle

External jugular vein

Clavicle

Occipital region

Mastoid process

Sternocleidomastoid muscle

Nuchal region

Trapezius muscle

Posterior cervical triangle

Acromion

Location of brachial plexus

PLATE **17**   THE POSTERIOR CERVICAL TRIANGLE

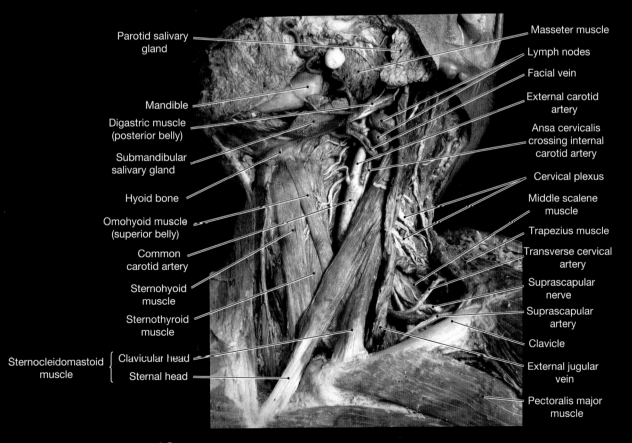

Parotid salivary gland

Mandible

Digastric muscle (posterior belly)

Submandibular salivary gland

Hyoid bone

Omohyoid muscle (superior belly)

Common carotid artery

Sternohyoid muscle

Sternothyroid muscle

Sternocleidomastoid muscle { Clavicular head
Sternal head

Masseter muscle

Lymph nodes

Facial vein

External carotid artery

Ansa cervicalis crossing internal carotid artery

Cervical plexus

Middle scalene muscle

Trapezius muscle

Transverse cervical artery

Suprascapular nerve

Suprascapular artery

Clavicle

External jugular vein

Pectoralis major muscle

PLATE **18a**   SUPERFICIAL STRUCTURES OF THE NECK, ANTEROLATERAL VIEW

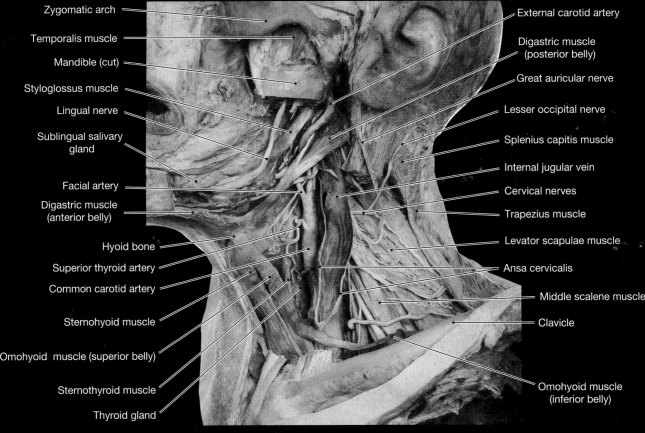

Zygomatic arch

Temporalis muscle

Mandible (cut)

Styloglossus muscle

Lingual nerve

Sublingual salivary gland

Facial artery

Digastric muscle (anterior belly)

Hyoid bone

Superior thyroid artery

Common carotid artery

Sternohyoid muscle

Omohyoid muscle (superior belly)

Sternothyroid muscle

Thyroid gland

External carotid artery

Digastric muscle (posterior belly)

Great auricular nerve

Lesser occipital nerve

Splenius capitis muscle

Internal jugular vein

Cervical nerves

Trapezius muscle

Levator scapulae muscle

Ansa cervicalis

Middle scalene muscle

Clavicle

Omohyoid muscle (inferior belly)

PLATE **18b** DEEPER STRUCTURES OF THE NECK, LATERAL VIEW

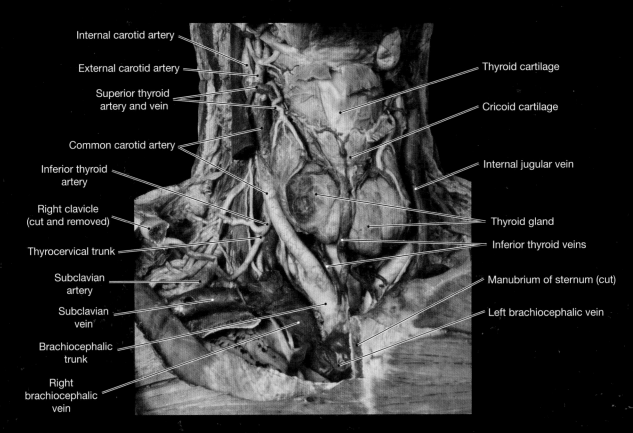

Internal carotid artery

External carotid artery

Superior thyroid artery and vein

Common carotid artery

Inferior thyroid artery

Right clavicle (cut and removed)

Thyrocervical trunk

Subclavian artery

Subclavian vein

Brachiocephalic trunk

Right brachiocephalic vein

Thyroid cartilage

Cricoid cartilage

Internal jugular vein

Thyroid gland

Inferior thyroid veins

Manubrium of sternum (cut)

Left brachiocephalic vein

PLATE **18c** DEEPER STRUCTURES OF THE NECK, ANTERIOR VIEW

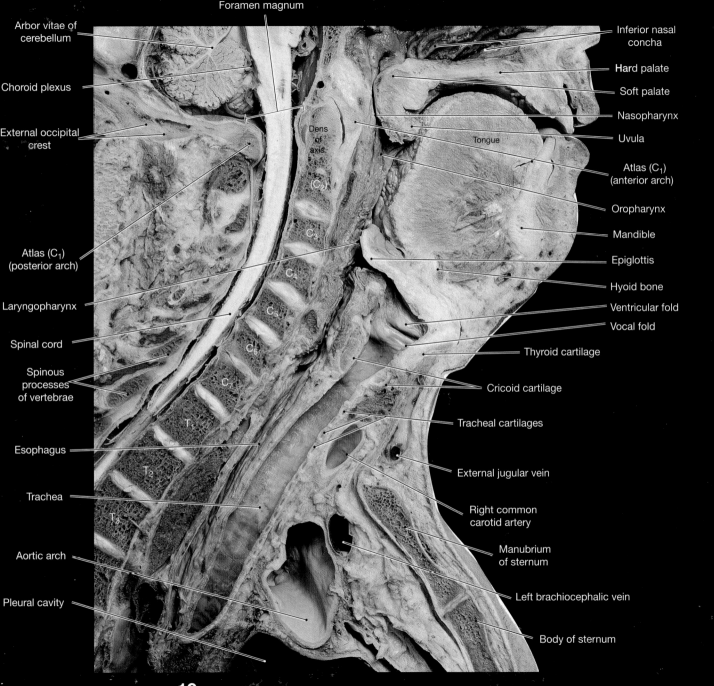

Foramen magnum

Arbor vitae of cerebellum

Choroid plexus

External occipital crest

Atlas (C₁) (posterior arch)

Laryngopharynx

Spinal cord

Spinous processes of vertebrae

Esophagus

Trachea

Aortic arch

Pleural cavity

Inferior nasal concha

Hard palate

Soft palate

Nasopharynx

Uvula

Atlas (C₁) (anterior arch)

Oropharynx

Mandible

Epiglottis

Hyoid bone

Ventricular fold

Vocal fold

Thyroid cartilage

Cricoid cartilage

Tracheal cartilages

External jugular vein

Right common carotid artery

Manubrium of sternum

Left brachiocephalic vein

Body of sternum

Dens of axis

(C₂)

C₃

C₄

C₅

C₆

C₇

T₁

T₂

T₃

Tongue

PLATE **19** MIDSAGITTAL SECTION THROUGH THE HEAD AND NECK

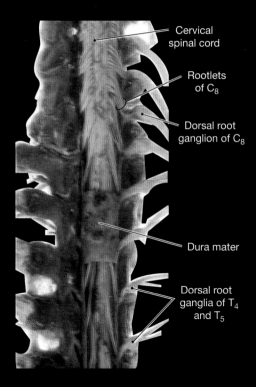

Cervical
spinal cord

Rootlets
of C_8

Dorsal root
ganglion of C_8

Dura mater

Dorsal root
ganglia of T_4
and T_5

PLATE 20a THE CERVICAL AND THORACIC
REGIONS OF THE SPINAL CORD, POSTERIOR VIEW

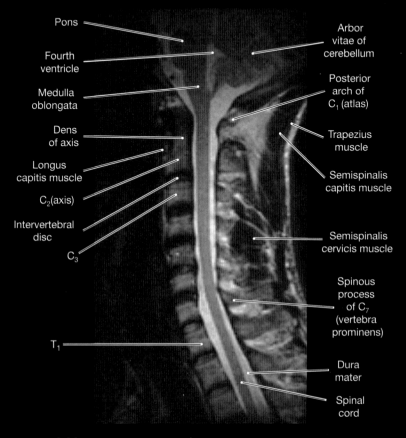

Pons

Fourth
ventricle

Medulla
oblongata

Dens
of axis

Longus
capitis muscle

C_2(axis)

Intervertebral
disc

C_3

T_1

Arbor
vitae of
cerebellum

Posterior
arch of
C_1 (atlas)

Trapezius
muscle

Semispinalis
capitis muscle

Semispinalis
cervicis muscle

Spinous
process
of C_7
(vertebra
prominens)

Dura
mater

Spinal
cord

PLATE 20b MRI SCAN OF CERVICAL REGION, SAGITTAL SECTION

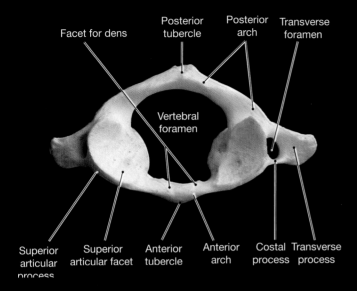

Facet for dens · Posterior tubercle · Posterior arch · Transverse foramen

Vertebral foramen

Superior articular process · Superior articular facet · Anterior tubercle · Anterior arch · Costal process · Transverse process

PLATE 21a ATLAS, SUPERIOR VIEW

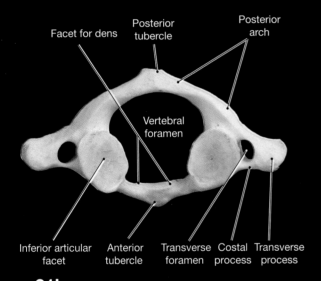

Facet for dens · Posterior tubercle · Posterior arch

Vertebral foramen

Inferior articular facet · Anterior tubercle · Transverse foramen · Costal process · Transverse process

PLATE 21b ATLAS, INFERIOR VIEW

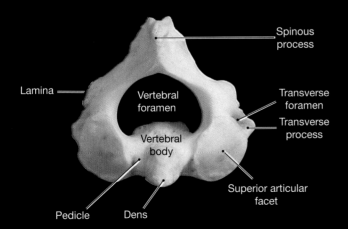

Spinous process

Lamina · Vertebral foramen · Transverse foramen · Transverse process

Vertebral body

Pedicle · Dens · Superior articular facet

PLATE 21c AXIS, SUPERIOR VIEW

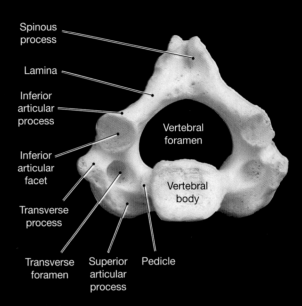

Spinous process

Lamina

Inferior articular process

Inferior articular facet

Transverse process

Vertebral foramen

Vertebral body

Transverse foramen · Superior articular process · Pedicle

PLATE 21d AXIS, INFERIOR VIEW

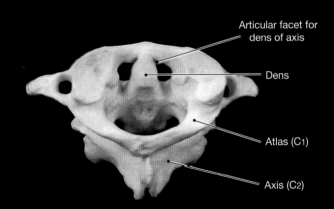

Articular facet for dens of axis

Dens

Atlas (C1)

Axis (C2)

PLATE 21e ARTICULATED ATLAS AND AXIS, SUPERIOR AND POSTERIOR VIEW

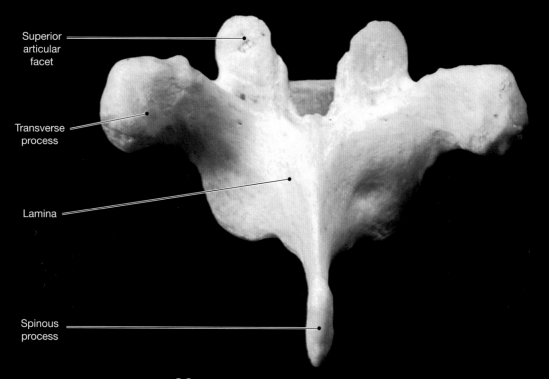

Superior articular facet

Transverse process

Lamina

Spinous process

PLATE **22a** THORACIC VERTEBRA, POSTERIOR VIEW

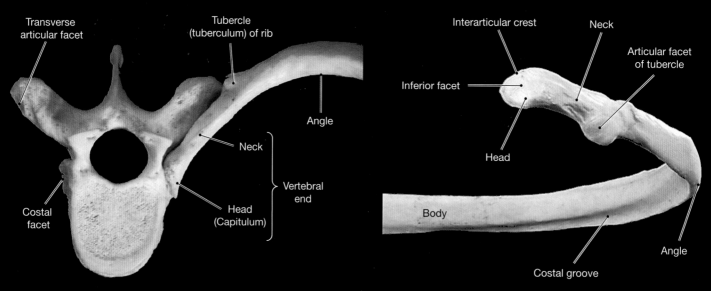

Transverse articular facet

Tubercle (tuberculum) of rib

Angle

Neck

Vertebral end

Head (Capitulum)

Costal facet

Interarticular crest

Neck

Articular facet of tubercle

Inferior facet

Head

Body

Angle

Costal groove

PLATE **22b** THORACIC VERTEBRA AND RIB, SUPERIOR VIEW

PLATE **22c** REPRESENTATIVE RIB, POSTERIOR VIEW

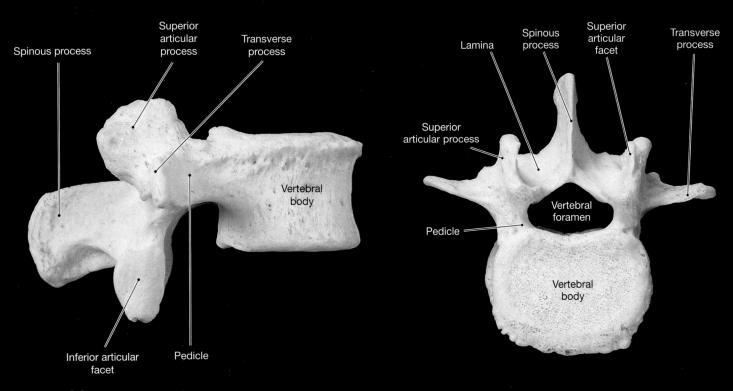

Spinous process

Superior articular process

Transverse process

Vertebral body

Inferior articular facet

Pedicle

PLATE 23a THE LUMBAR VERTEBRAE, LATERAL VIEW

Lamina

Spinous process

Superior articular facet

Transverse process

Superior articular process

Vertebral foramen

Pedicle

Vertebral body

PLATE 23b THE LUMBAR VERTEBRAE, SUPERIOR VIEW

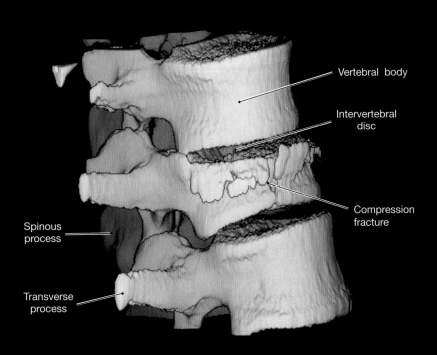

Vertebral body

Intervertebral disc

Compression fracture

Spinous process

Transverse process

PLATE 23c 3-DIMENSIONAL CT SCAN SHOWING A FRACTURE OF THE BODY OF A LUMBAR VERTEBRA

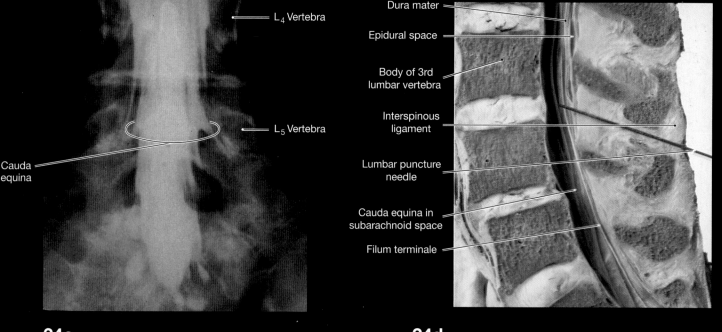

L₄ Vertebra

L₅ Vertebra

Cauda
equina

Dura mater

Epidural space

Body of 3rd
lumbar vertebra

Interspinous
ligament

Lumbar puncture
needle

Cauda equina in
subarachnoid space

Filum terminale

PLATE 24c X-RAY OF THE CAUDA EQUINA WITH A
CONTRAST MEDIUM IN THE SUBARACHNOID SPACE

PLATE 24d LUMBAR PUNCTURE POSITIONING, SEEN IN A
SAGITTAL SECTION

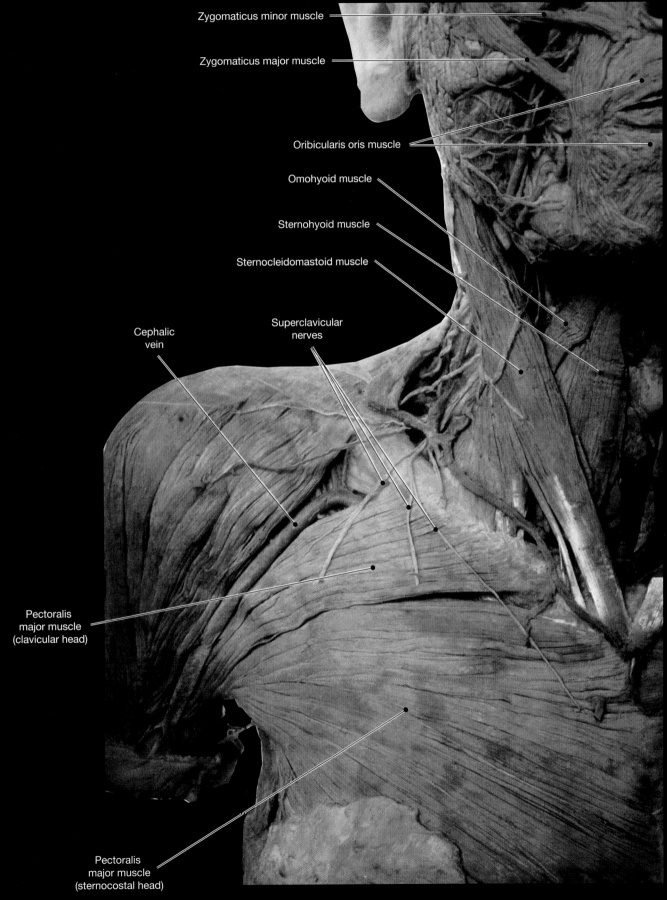

Zygomaticus minor muscle

Zygomaticus major muscle

Oribicularis oris muscle

Omohyoid muscle

Sternohyoid muscle

Sternocleidomastoid muscle

Cephalic
vein

Superclavicular
nerves

Pectoralis
major muscle
(clavicular head)

Pectoralis
major muscle
(sternocostal head)

PLATE 25 SHOULDER AND NECK, ANTERIOR VIEW

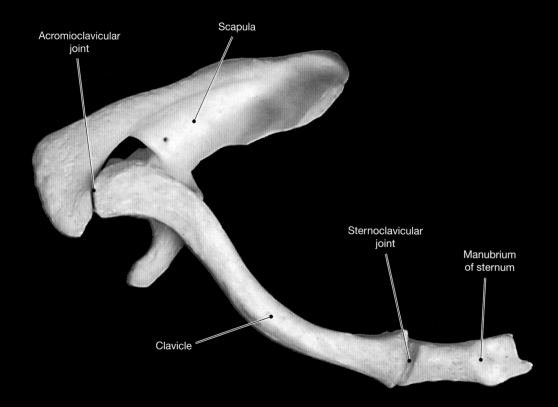

Acromioclavicular joint

Scapula

Sternoclavicular joint

Manubrium of sternum

Clavicle

PLATE 26a   BONES OF THE RIGHT PECTORAL GIRDLE, SUPERIOR VIEW

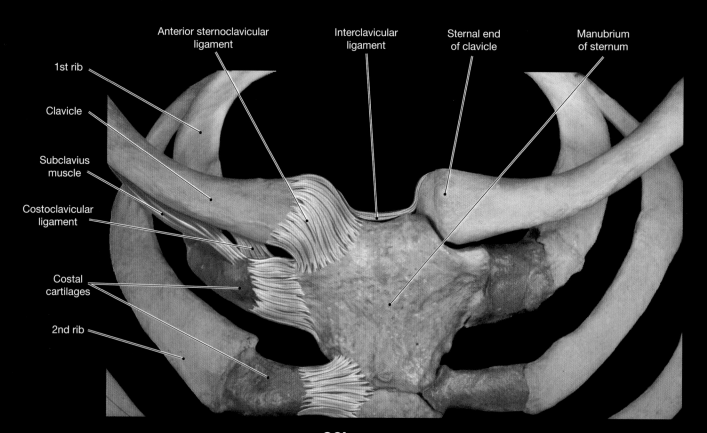

Anterior sternoclavicular ligament

Interclavicular ligament

Sternal end of clavicle

Manubrium of sternum

1st rib

Clavicle

Subclavius muscle

Costoclavicular ligament

Costal cartilages

2nd rib

PLATE 26b   STERNOCLAVICULAR JOINT

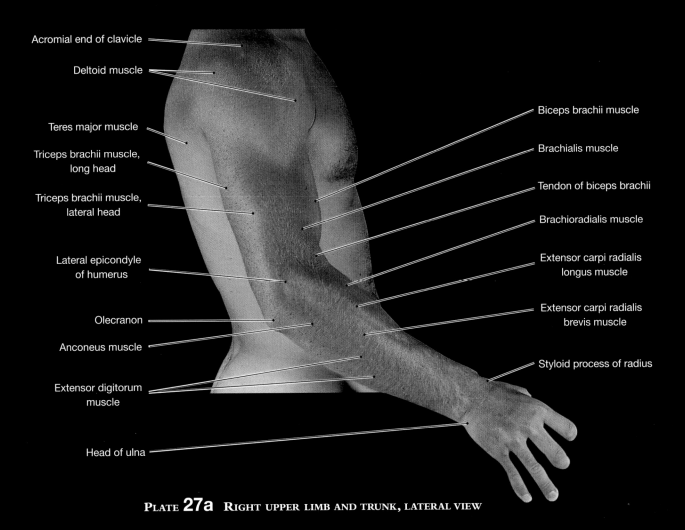

Acromial end of clavicle

Deltoid muscle

Teres major muscle

Triceps brachii muscle, long head

Triceps brachii muscle, lateral head

Lateral epicondyle of humerus

Olecranon

Anconeus muscle

Extensor digitorum muscle

Head of ulna

Biceps brachii muscle

Brachialis muscle

Tendon of biceps brachii

Brachioradialis muscle

Extensor carpi radialis longus muscle

Extensor carpi radialis brevis muscle

Styloid process of radius

PLATE **27a** RIGHT UPPER LIMB AND TRUNK, LATERAL VIEW

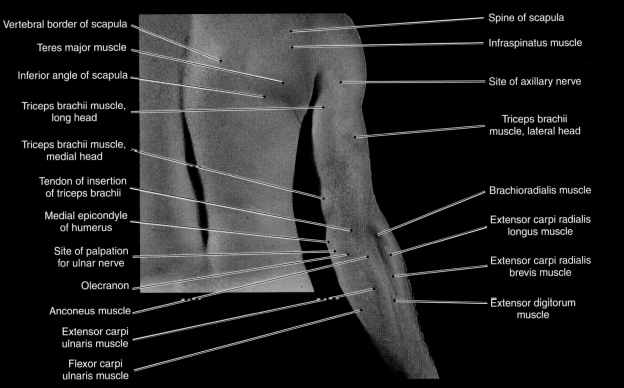

Vertebral border of scapula

Teres major muscle

Inferior angle of scapula

Triceps brachii muscle, long head

Triceps brachii muscle, medial head

Tendon of insertion of triceps brachii

Medial epicondyle of humerus

Site of palpation for ulnar nerve

Olecranon

Anconeus muscle

Extensor carpi ulnaris muscle

Flexor carpi ulnaris muscle

Spine of scapula

Infraspinatus muscle

Site of axillary nerve

Triceps brachii muscle, lateral head

Brachioradialis muscle

Extensor carpi radialis longus muscle

Extensor carpi radialis brevis muscle

Extensor digitorum muscle

PLATE **27b** RIGHT UPPER LIMB AND TRUNK, POSTERIOR VIEW

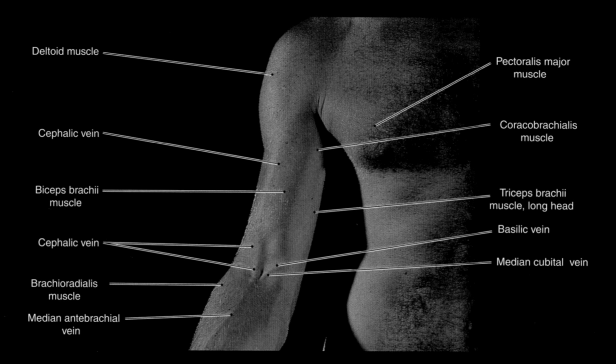

Deltoid muscle

Cephalic vein

Biceps brachii
muscle

Cephalic vein

Brachioradialis
muscle

Median antebrachial
vein

Pectoralis major
muscle

Coracobrachialis
muscle

Triceps brachii
muscle, long head

Basilic vein

Median cubital vein

PLATE 27c RIGHT ARM AND TRUNK, ANTERIOR VIEW

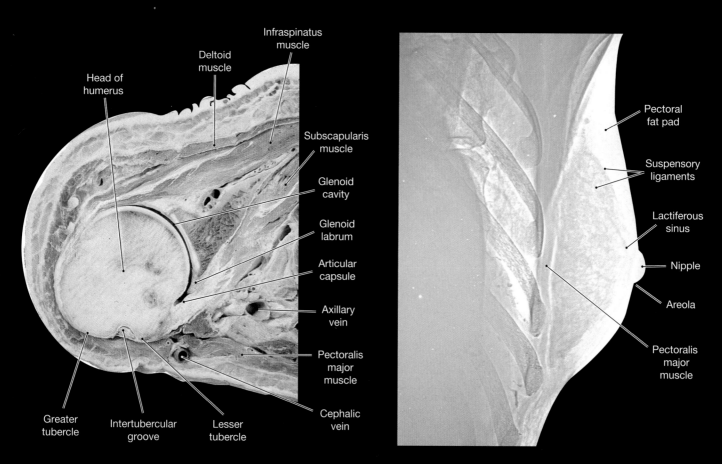

Head of
humerus

Deltoid
muscle

Infraspinatus
muscle

Subscapularis
muscle

Glenoid
cavity

Glenoid
labrum

Articular
capsule

Axillary
vein

Pectoralis
major
muscle

Cephalic
vein

Greater
tubercle

Intertubercular
groove

Lesser
tubercle

Pectoral
fat pad

Suspensory
ligaments

Lactiferous
sinus

Nipple

Areola

Pectoralis
major
muscle

**PLATE 27d HORIZONTAL SECTION THROUGH THE RIGHT
SHOULDER**

PLATE 28 XEROMAMMOGRAM

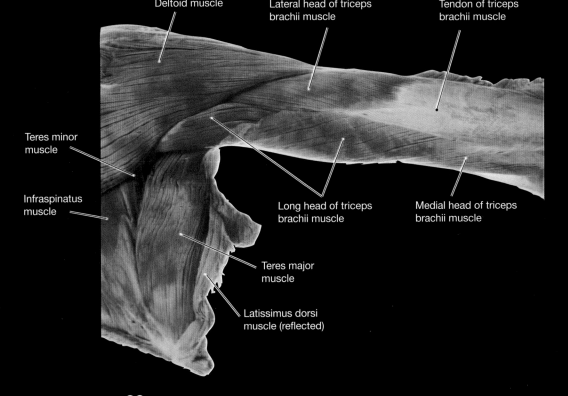

Deltoid muscle

Lateral head of triceps brachii muscle

Tendon of triceps brachii muscle

Teres minor muscle

Infraspinatus muscle

Long head of triceps brachii muscle

Medial head of triceps brachii muscle

Teres major muscle

Latissimus dorsi muscle (reflected)

PLATE **29a** SUPERFICIAL DISSECTION OF RIGHT SHOULDER, POSTERIOR VIEW

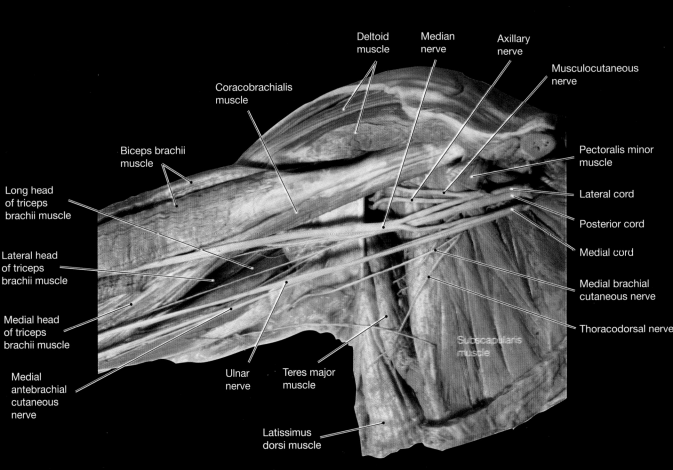

Deltoid muscle

Median nerve

Axillary nerve

Musculocutaneous nerve

Coracobrachialis muscle

Biceps brachii muscle

Pectoralis minor muscle

Long head of triceps brachii muscle

Lateral cord

Posterior cord

Lateral head of triceps brachii muscle

Medial cord

Medial brachial cutaneous nerve

Medial head of triceps brachii muscle

Thoracodorsal nerve

Subscapularis muscle

Medial antebrachial cutaneous nerve

Ulnar nerve

Teres major muscle

Latissimus dorsi muscle

PLATE **29b** DEEP DISSECTION OF THE RIGHT BRACHIAL PLEXUS

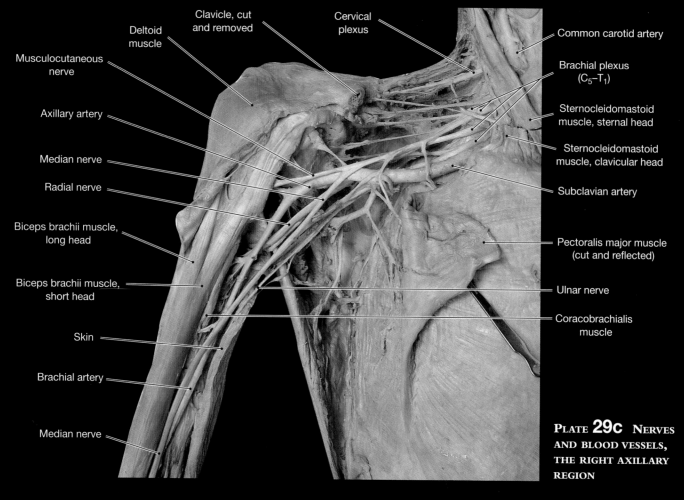

Clavicle, cut and removed

Cervical plexus

Deltoid muscle

Musculocutaneous nerve

Axillary artery

Median nerve

Radial nerve

Biceps brachii muscle, long head

Biceps brachii muscle, short head

Skin

Brachial artery

Median nerve

Common carotid artery

Brachial plexus (C$_5$–T$_1$)

Sternocleidomastoid muscle, sternal head

Sternocleidomastoid muscle, clavicular head

Subclavian artery

Pectoralis major muscle (cut and reflected)

Ulnar nerve

Coracobrachialis muscle

PLATE 29c NERVES AND BLOOD VESSELS, THE RIGHT AXILLARY REGION

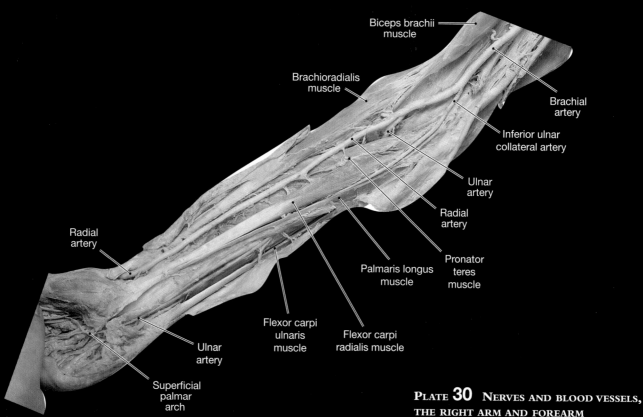

Biceps brachii muscle

Brachioradialis muscle

Radial artery

Ulnar artery

Superficial palmar arch

Flexor carpi ulnaris muscle

Flexor carpi radialis muscle

Palmaris longus muscle

Pronator teres muscle

Radial artery

Ulnar artery

Inferior ulnar collateral artery

Brachial artery

PLATE 30 NERVES AND BLOOD VESSELS, THE RIGHT ARM AND FOREARM

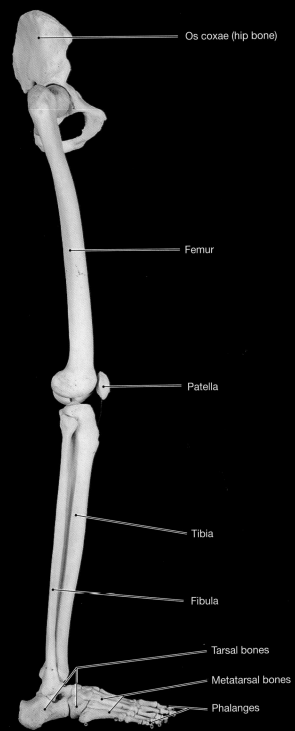

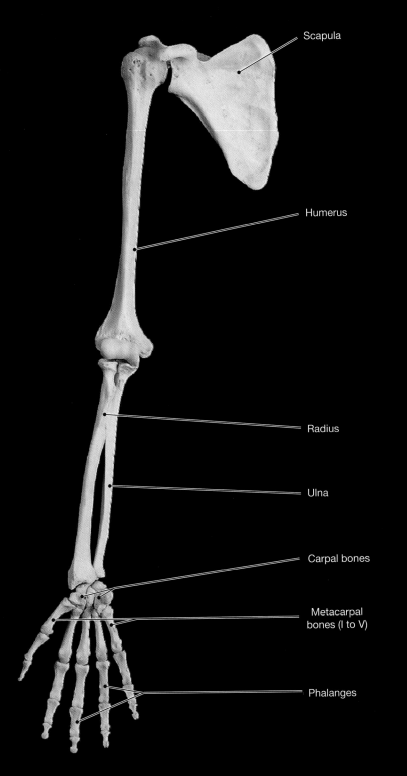

Scapula

Os coxae (hip bone)

Humerus

Femur

Radius

Patella

Ulna

Carpal bones

Tibia

Metacarpal
bones (I to V)

Fibula

Tarsal bones

Metatarsal bones

Phalanges

Phalanges

**PLATE 31 BONES OF THE RIGHT UPPER LIMB, ANTERIOR
VIEW**

**PLATE 32 BONES OF THE RIGHT LOWER LIMB, LATERAL
VIEW**

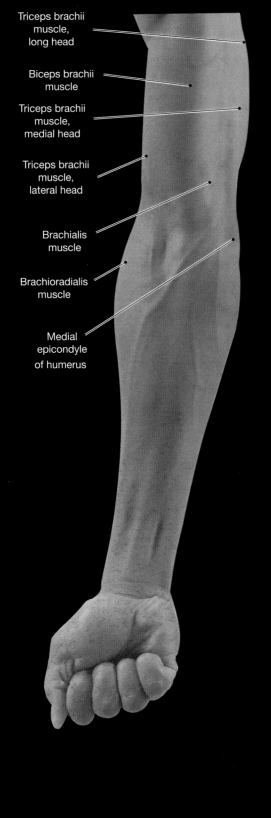

Triceps brachii
muscle,
long head

Biceps brachii
muscle

Triceps brachii
muscle,
medial head

Triceps brachii
muscle,
lateral head

Brachialis
muscle

Brachioradialis
muscle

Medial
epicondyle
of humerus

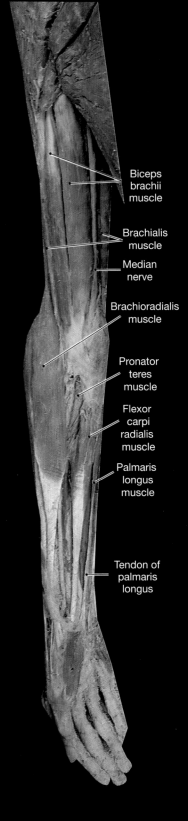

Biceps
brachii
muscle

Brachialis
muscle

Median
nerve

Brachioradialis
muscle

Pronator
teres
muscle

Flexor
carpi
radialis
muscle

Palmaris
longus
muscle

Tendon of
palmaris
longus

PLATE 33a THE RIGHT UPPER LIMB, ANTERIOR SURFACE,
MUSCLES

PLATE 33b THE RIGHT UPPER LIMB, ANTERIOR VIEW,
SUPERFICIAL DISSECTION

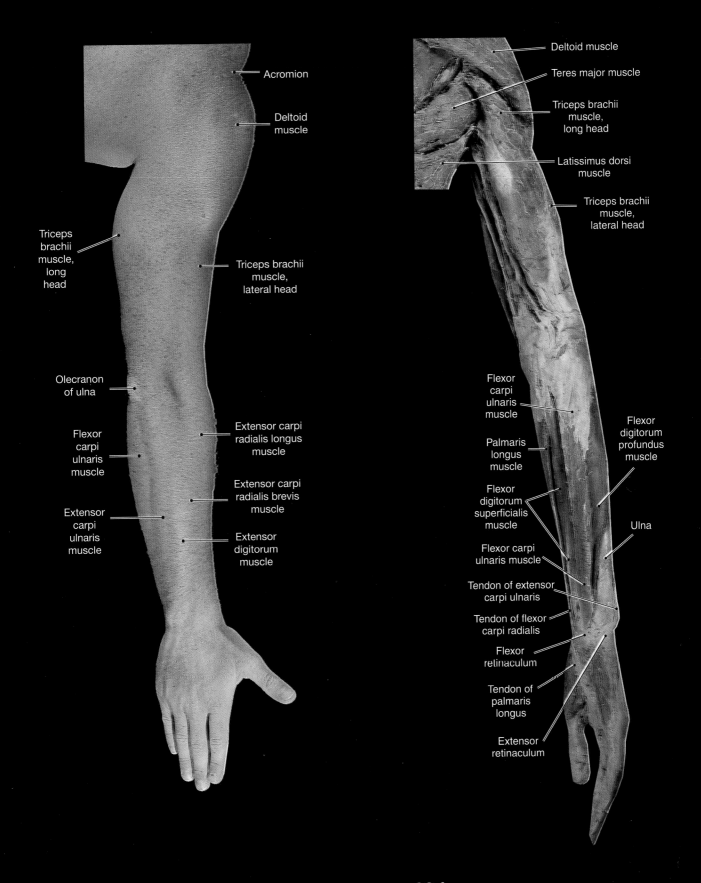

Acromion

Deltoid
muscle

Triceps
brachii
muscle,
long
head

Triceps brachii
muscle,
lateral head

Olecranon
of ulna

Flexor
carpi
ulnaris
muscle

Extensor carpi
radialis longus
muscle

Extensor carpi
radialis brevis
muscle

Extensor
carpi
ulnaris
muscle

Extensor
digitorum
muscle

Deltoid muscle

Teres major muscle

Triceps brachii
muscle,
long head

Latissimus dorsi
muscle

Triceps brachii
muscle,
lateral head

Flexor
carpi
ulnaris
muscle

Flexor
digitorum
profundus
muscle

Palmaris
longus
muscle

Flexor
digitorum
superficialis
muscle

Flexor carpi
ulnaris muscle

Ulna

Tendon of extensor
carpi ulnaris

Tendon of flexor
carpi radialis

Flexor
retinaculum

Tendon of
palmaris
longus

Extensor
retinaculum

PLATE **33c** THE RIGHT UPPER LIMB, POSTERIOR SURFACE,
LANDMARKS

PLATE **33d** THE RIGHT UPPER LIMB, POSTERIOR VIEW,
SUPERFICIAL DISSECTION

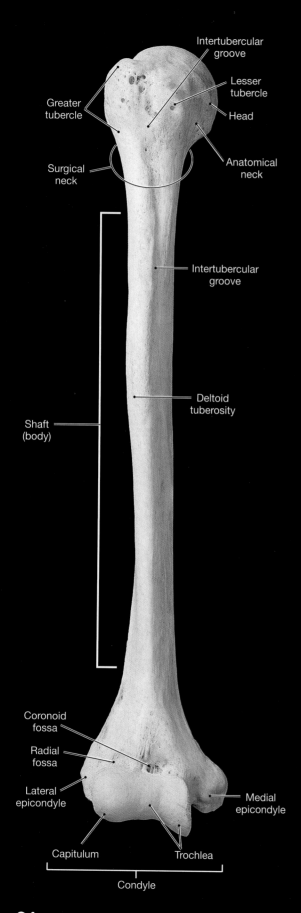

Intertubercular groove

Lesser tubercle

Greater tubercle

Head

Surgical neck

Anatomical neck

Intertubercular groove

Deltoid tuberosity

Shaft (body)

Coronoid fossa

Radial fossa

Lateral epicondyle

Medial epicondyle

Capitulum

Trochlea

Condyle

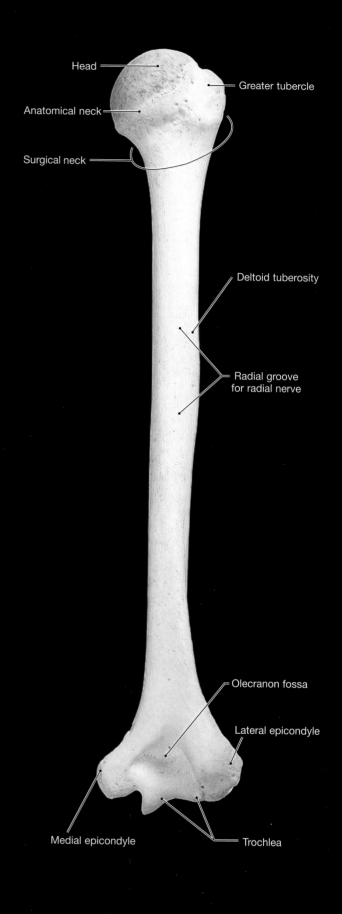

Head

Greater tubercle

Anatomical neck

Surgical neck

Deltoid tuberosity

Radial groove for radial nerve

Olecranon fossa

Lateral epicondyle

Medial epicondyle

Trochlea

PLATE **34a** RIGHT HUMERUS, ANTERIOR VIEW

PLATE **34b** RIGHT HUMERUS, POSTERIOR VIEW

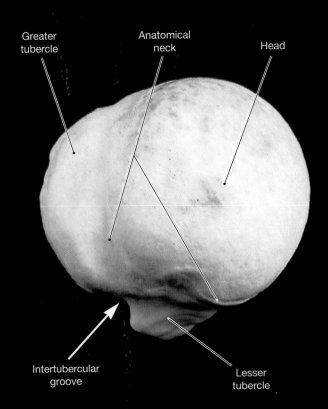

Greater tubercle

Anatomical neck

Head

Intertubercular groove

Lesser tubercle

PLATE 34c PROXIMAL END OF RIGHT HUMERUS, SUPERIOR VIEW

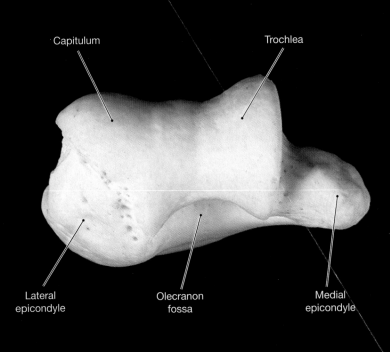

Capitulum

Trochlea

Lateral epicondyle

Olecranon fossa

Medial epicondyle

PLATE 34d DISTAL END OF RIGHT HUMERUS, INFERIOR VIEW

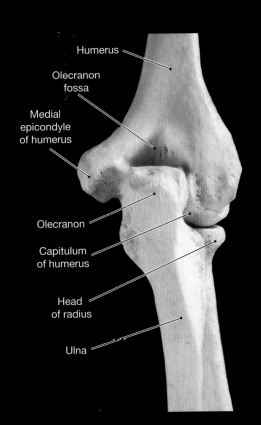

Humerus

Olecranon fossa

Medial epicondyle of humerus

Olecranon

Capitulum of humerus

Head of radius

Ulna

PLATE 35a RIGHT ELBOW JOINT, POSTERIOR VIEW

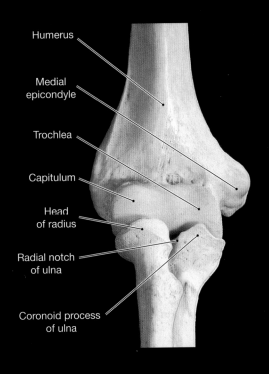

Humerus

Medial epicondyle

Trochlea

Capitulum

Head of radius

Radial notch of ulna

Coronoid process of ulna

PLATE 35b RIGHT ELBOW JOINT, ANTERIOR VIEW

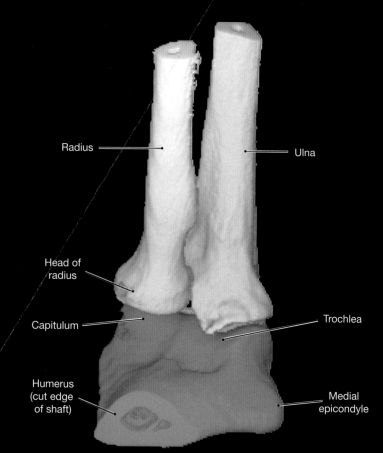

Radius

Ulna

Head of radius

Capitulum

Trochlea

Humerus (cut edge of shaft)

Medial epicondyle

PLATE 35c 3-DIMENSIONAL CT SCAN OF THE RIGHT
ELBOW JOINT, SUPERIOR VIEW

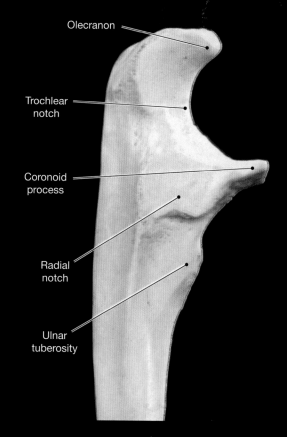

Medial epicondyle of humerus

Capitulum of humerus

Trochlea of humerus

Annular ligament

Articular capsule

Head of radius

Coronoid process of ulna

Radial notch of ulna

Trochlear notch of ulna

Olecranon

PLATE 35e ARTICULAR SURFACES WITHIN THE RIGHT
ELBOW JOINT

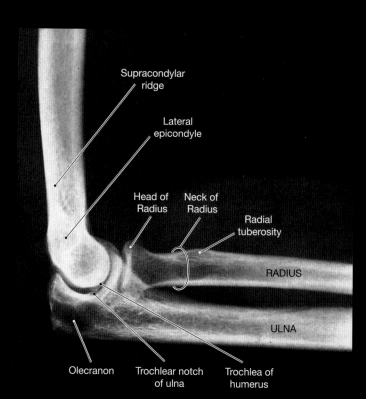

Supracondylar ridge

Lateral epicondyle

Head of Radius

Neck of Radius

Radial tuberosity

RADIUS

ULNA

Olecranon

Trochlear notch of ulna

Trochlea of humerus

PLATE 35d X-RAY OF THE ELBOW JOINT, MEDIAL-LATERAL
PROJECTION

Olecranon

Trochlear notch

Coronoid process

Radial notch

Ulnar tuberosity

PLATE 35f RIGHT ULNA, LATERAL VIEW

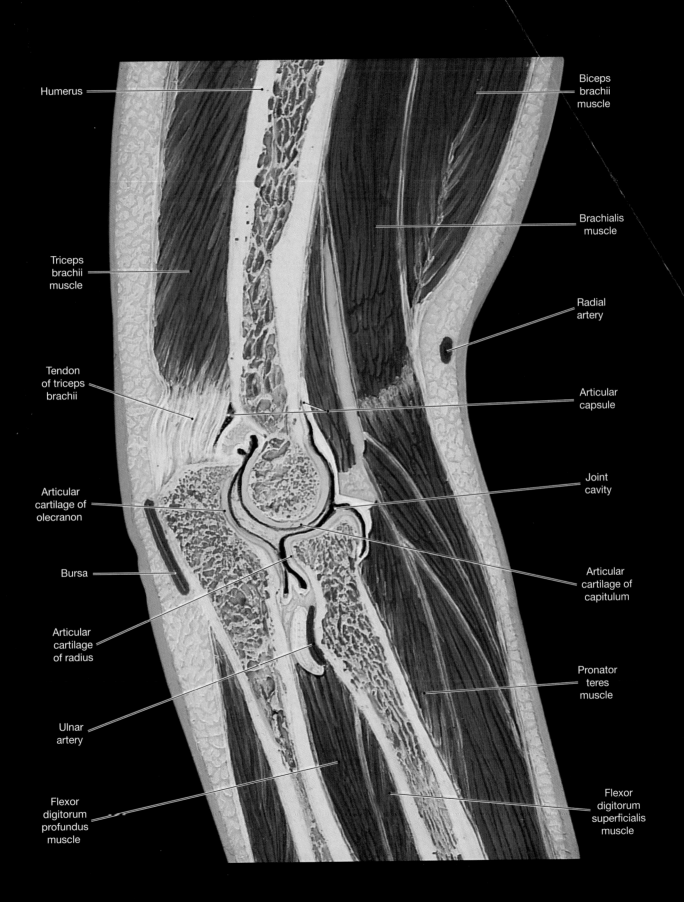

Humerus

Biceps brachii muscle

Brachialis muscle

Triceps brachii muscle

Radial artery

Tendon of triceps brachii

Articular capsule

Articular cartilage of olecranon

Joint cavity

Bursa

Articular cartilage of capitulum

Articular cartilage of radius

Pronator teres muscle

Ulnar artery

Flexor digitorum profundus muscle

Flexor digitorum superficialis muscle

PLATE **35g** THE ELBOW, OBLIQUE SECTION—MODEL

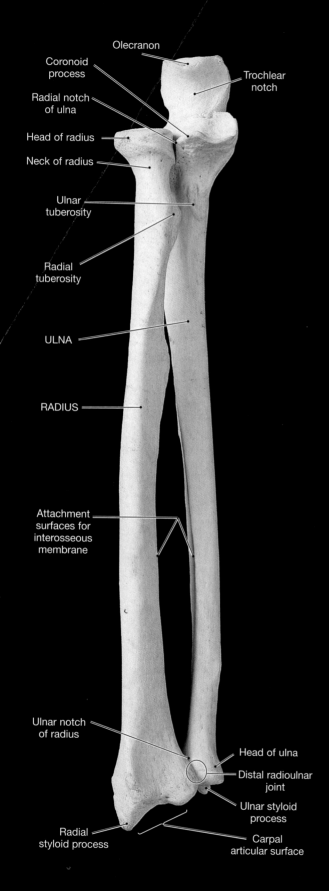

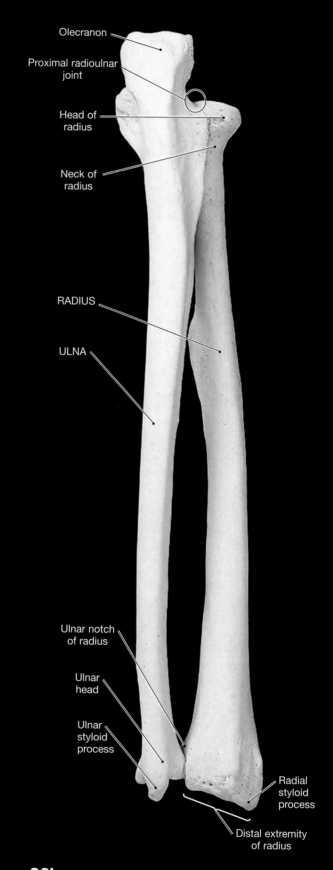

Olecranon

Coronoid
process

Radial notch
of ulna

Head of radius

Neck of radius

Ulnar
tuberosity

Radial
tuberosity

ULNA

RADIUS

Attachment
surfaces for
interosseous
membrane

Ulnar notch
of radius

Radial
styloid process

Trochlear
notch

Head of ulna

Distal radioulnar
joint

Ulnar styloid
process

Carpal
articular surface

Olecranon

Proximal radioulnar
joint

Head of
radius

Neck of
radius

RADIUS

ULNA

Ulnar notch
of radius

Ulnar
head

Ulnar
styloid
process

Radial
styloid
process

Distal extremity
of radius

PLATE **36a** RIGHT RADIUS AND ULNA, ANTERIOR VIEW   PLATE **36b** RIGHT RADIUS AND ULNA, POSTERIOR VIEW

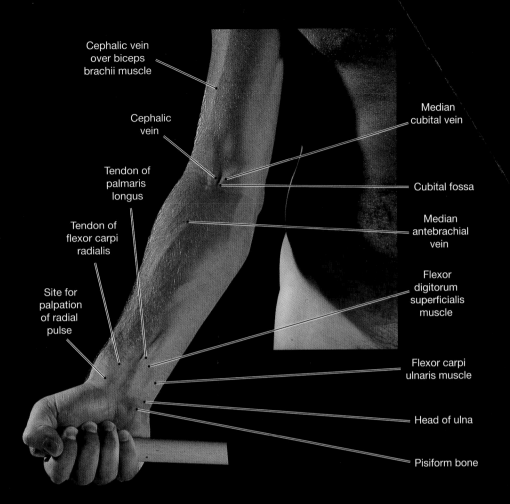

Cephalic vein
over biceps
brachii muscle

Cephalic
vein

Tendon of
palmaris
longus

Tendon of
flexor carpi
radialis

Site for
palpation
of radial
pulse

Median
cubital vein

Cubital fossa

Median
antebrachial
vein

Flexor
digitorum
superficialis
muscle

Flexor carpi
ulnaris muscle

Head of ulna

Pisiform bone

PLATE **37a** THE RIGHT UPPER LIMB, ANTERIOR SURFACE, LANDMARKS

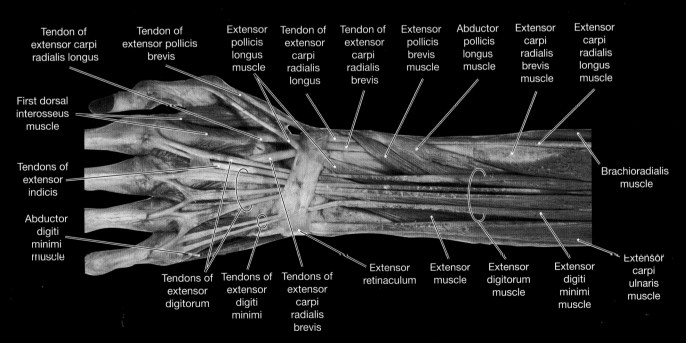

Tendon of
extensor carpi
radialis longus

Tendon of
extensor pollicis
brevis

Extensor
pollicis
longus
muscle

Tendon of
extensor
carpi
radialis
longus

Tendon of
extensor
carpi
radialis
brevis

Extensor
pollicis
brevis
muscle

Abductor
pollicis
longus
muscle

Extensor
carpi
radialis
brevis
muscle

Extensor
carpi
radialis
longus
muscle

First dorsal
interosseus
muscle

Tendons of
extensor
indicis

Abductor
digiti
minimi
muscle

Brachioradialis
muscle

Tendons of
extensor
digitorum

Tendons of
extensor
digiti
minimi

Tendons of
extensor
carpi
radialis
brevis

Extensor
retinaculum

Extensor
muscle

Extensor
digitorum
muscle

Extensor
digiti
minimi
muscle

Extensor
carpi
ulnaris
muscle

PLATE **37b** SUPERFICIAL DISSECTION OF RIGHT FOREARM AND HAND, POSTERIOR VIEW

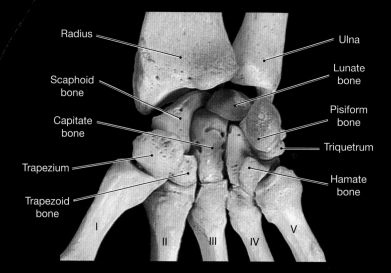

Radius

Scaphoid
bone

Capitate
bone

Trapezium

Trapezoid
bone

Ulna

Lunate
bone

Pisiform
bone

Triquetrum

Hamate
bone

I   II   III   IV   V

PLATE 38a **BONES OF THE RIGHT WRIST, ANTERIOR VIEW**

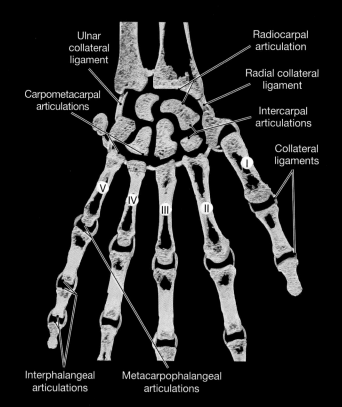

Ulnar
collateral
ligament

Carpometacarpal
articulations

Radiocarpal
articulation

Radial collateral
ligament

Intercarpal
articulations

Collateral
ligaments

V   IV   III   II   I

Interphalangeal
articulations

Metacarpophalangeal
articulations

PLATE 38b **JOINTS OF THE RIGHT WRIST, CORONAL
SECTION**

PLATE **38c** THE RIGHT HAND,
POSTERIOR VIEW—MODEL

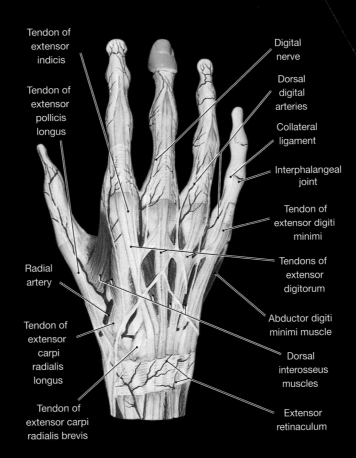

Tendon of extensor indicis

Tendon of extensor pollicis longus

Digital nerve

Dorsal digital arteries

Collateral ligament

Interphalangeal joint

Tendon of extensor digiti minimi

Radial artery

Tendon of extensor carpi radialis longus

Tendon of extensor carpi radialis brevis

Tendons of extensor digitorum

Abductor digiti minimi muscle

Dorsal interosseus muscles

Extensor retinaculum

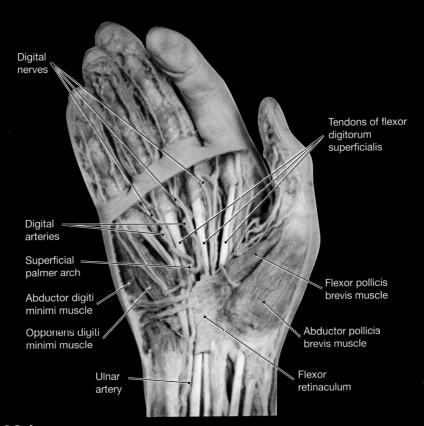

Digital nerves

Tendons of flexor digitorum superficialis

Digital arteries

Superficial palmer arch

Abductor digiti minimi muscle

Opponens digiti minimi muscle

Ulnar artery

Flexor pollicis brevis muscle

Abductor pollicis brevis muscle

Flexor retinaculum

PLATE **38d** SUPERFICIAL DISSECTION OF THE RIGHT WRIST AND HAND, ANTERIOR VIEW

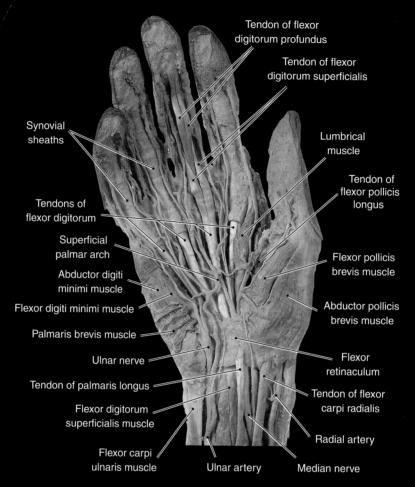

Tendon of flexor
digitorum profundus

Tendon of flexor
digitorum superficialis

Synovial
sheaths

Lumbrical
muscle

Tendon of
flexor pollicis
longus

Tendons of
flexor digitorum

Superficial
palmar arch

Abductor digiti
minimi muscle

Flexor digiti minimi muscle

Palmaris brevis muscle

Ulnar nerve

Tendon of palmaris longus

Flexor digitorum
superficialis muscle

Flexor carpi
ulnaris muscle

Ulnar artery

Flexor pollicis
brevis muscle

Abductor pollicis
brevis muscle

Flexor
retinaculum

Tendon of flexor
carpi radialis

Radial artery

Median nerve

PLATE **38e** THE RIGHT HAND, ANTERIOR VIEW, SUPERFICIAL DISSECTION

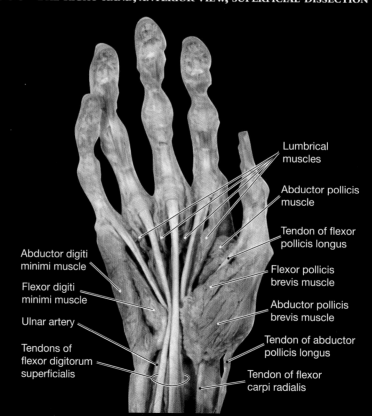

Lumbrical
muscles

Abductor pollicis
muscle

Tendon of flexor
pollicis longus

Flexor pollicis
brevis muscle

Abductor pollicis
brevis muscle

Tendon of abductor
pollicis longus

Tendon of flexor
carpi radialis

Abductor digiti
minimi muscle

Flexor digiti
minimi muscle

Ulnar artery

Tendons of
flexor digitorum
superficialis

PLATE **38f** FLEXOR TENDONS OF RIGHT WRIST AND HAND

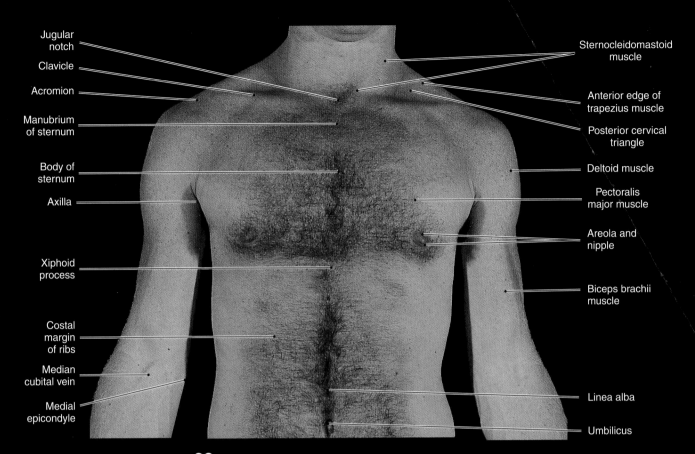

Jugular notch

Clavicle

Acromion

Manubrium of sternum

Body of sternum

Axilla

Xiphoid process

Costal margin of ribs

Median cubital vein

Medial epicondyle

Sternocleidomastoid muscle

Anterior edge of trapezius muscle

Posterior cervical triangle

Deltoid muscle

Pectoralis major muscle

Areola and nipple

Biceps brachii muscle

Linea alba

Umbilicus

PLATE **39a** SURFACE ANATOMY OF THE TRUNK, ANTERIOR VIEW

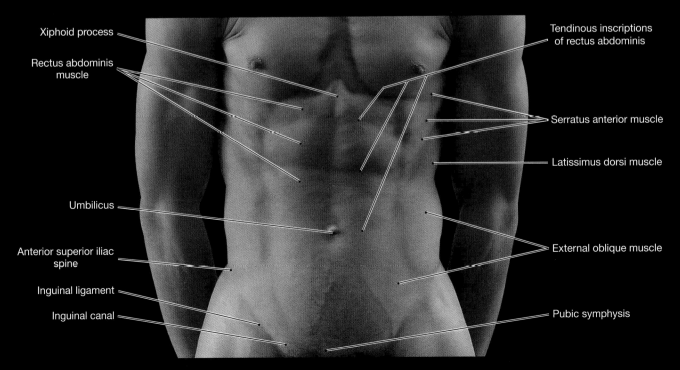

Xiphoid process

Rectus abdominis muscle

Umbilicus

Anterior superior iliac spine

Inguinal ligament

Inguinal canal

Tendinous inscriptions of rectus abdominis

Serratus anterior muscle

Latissimus dorsi muscle

External oblique muscle

Pubic symphysis

PLATE **39b** SURFACE ANATOMY OF THE ABDOMEN, ANTERIOR VIEW

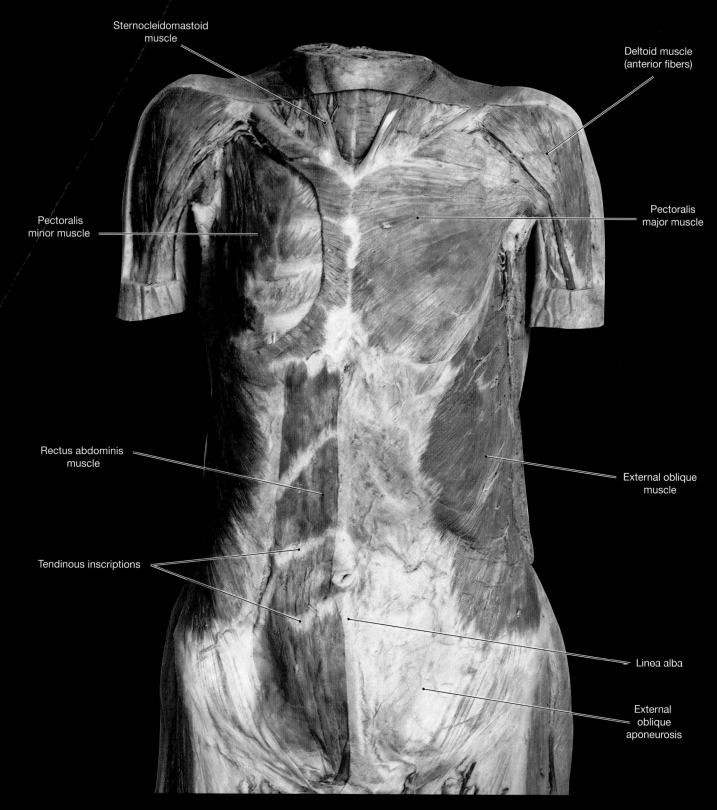

Sternocleidomastoid muscle

Deltoid muscle (anterior fibers)

Pectoralis minor muscle

Pectoralis major muscle

Rectus abdominis muscle

External oblique muscle

Tendinous inscriptions

Linea alba

External oblique aponeurosis

PLATE **39c** TRUNK, ANTERIOR VIEW

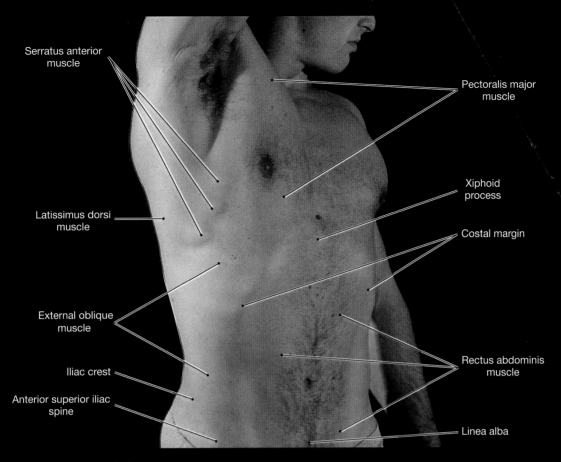

Serratus anterior
muscle

Pectoralis major
muscle

Xiphoid
process

Latissimus dorsi
muscle

Costal margin

External oblique
muscle

Rectus abdominis
muscle

Iliac crest

Anterior superior iliac
spine

Linea alba

PLATE **39d** SURFACE ANATOMY OF THE ABDOMEN, ANTEROLATERAL VIEW

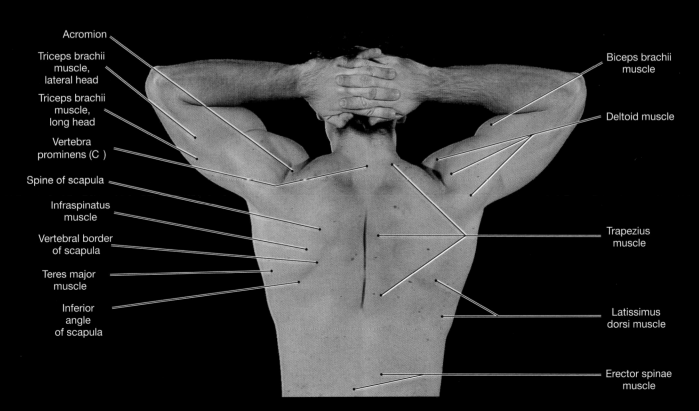

Acromion

Biceps brachii
muscle

Triceps brachii
muscle,
lateral head

Deltoid muscle

Triceps brachii
muscle,
long head

Vertebra
prominens (C)

Spine of scapula

Infraspinatus
muscle

Trapezius
muscle

Vertebral border
of scapula

Teres major
muscle

Inferior
angle
of scapula

Latissimus
dorsi muscle

Erector spinae
muscle

PLATE **40a** SURFACE ANATOMY OF THE TRUNK, POSTERIOR VIEW

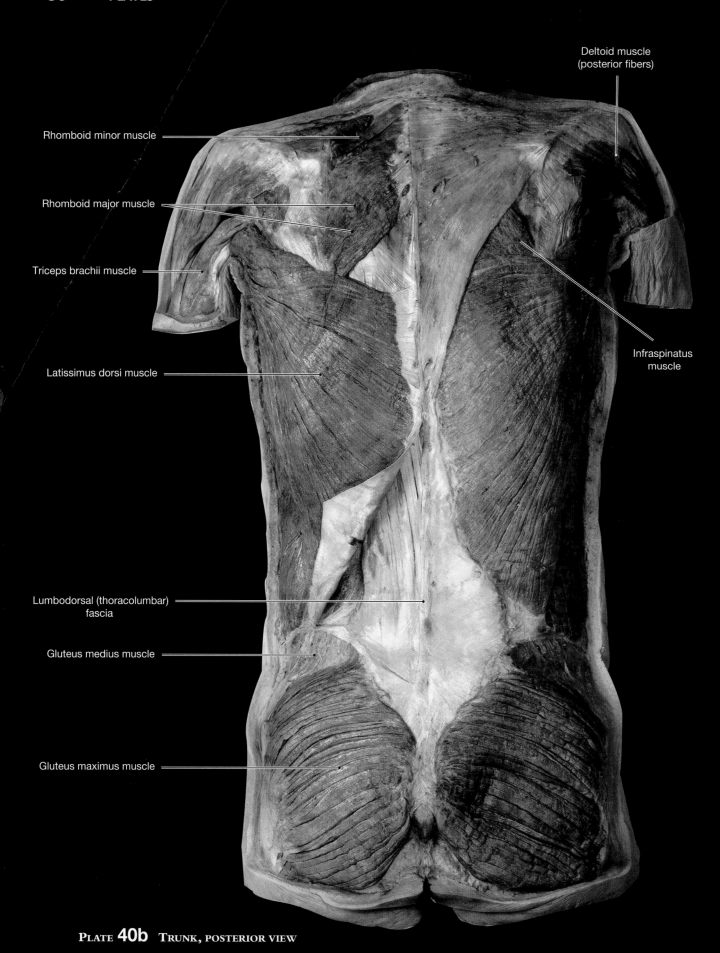

Deltoid muscle
(posterior fibers)

Rhomboid minor muscle

Rhomboid major muscle

Triceps brachii muscle

Infraspinatus
muscle

Latissimus dorsi muscle

Lumbodorsal (thoracolumbar)
fascia

Gluteus medius muscle

Gluteus maximus muscle

PLATE **40b** TRUNK, POSTERIOR VIEW

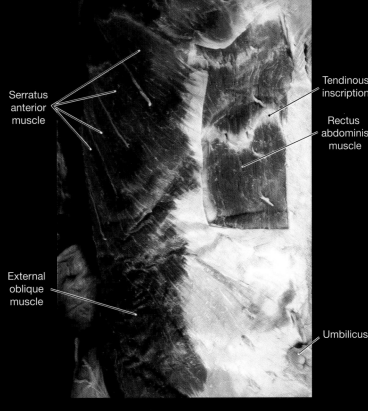

Serratus anterior muscle

Tendinous inscription

Rectus abdominis muscle

External oblique muscle

Umbilicus

PLATE **41a** ABDOMINAL WALL, ANTERIOR VIEW

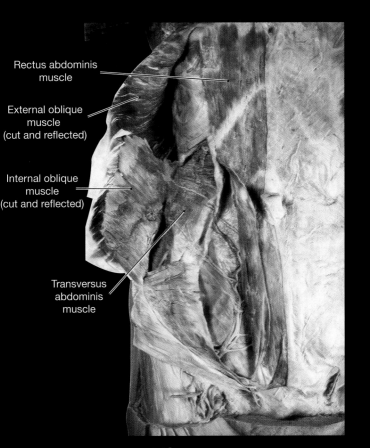

Rectus abdominis muscle

External oblique muscle (cut and reflected)

Internal oblique muscle (cut and reflected)

Transversus abdominis muscle

PLATE **41b** ABDOMINAL MUSCLES

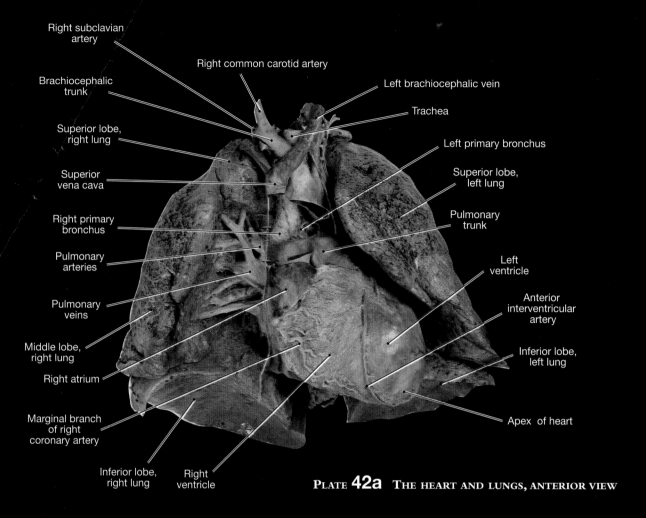

Right subclavian artery

Right common carotid artery

Brachiocephalic trunk

Left brachiocephalic vein

Superior lobe, right lung

Trachea

Superior vena cava

Left primary bronchus

Right primary bronchus

Superior lobe, left lung

Pulmonary arteries

Pulmonary trunk

Pulmonary veins

Left ventricle

Middle lobe, right lung

Anterior interventricular artery

Right atrium

Inferior lobe, left lung

Marginal branch of right coronary artery

Apex of heart

Inferior lobe, right lung

Right ventricle

PLATE **42a** THE HEART AND LUNGS, ANTERIOR VIEW

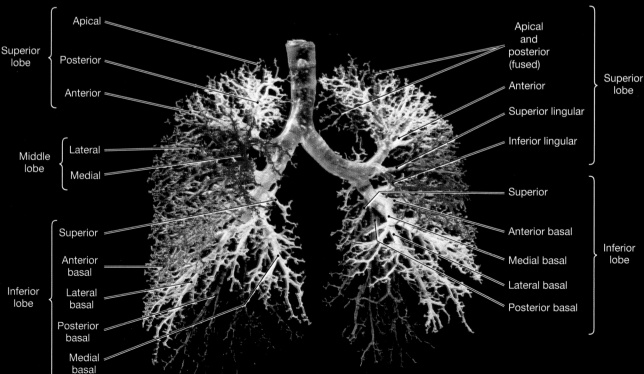

Apical

Apical and posterior (fused)

Superior lobe

Posterior

Anterior

Anterior

Superior lingular

Superior lobe

Middle lobe

Lateral

Inferior lingular

Medial

Superior

Inferior lobe

Superior

Anterior basal

Anterior basal

Medial basal

Lateral basal

Lateral basal

Inferior lobe

Posterior basal

Posterior basal

Medial basal

PLATE **42b** COLOR-CODED CORROSION CAST OF THE BRONCHIAL TREE, ANTERIOR VIEW

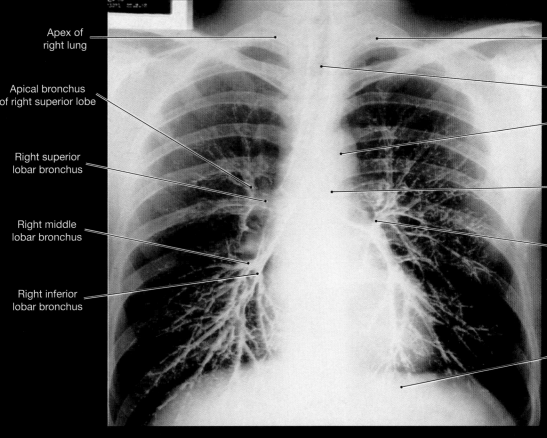

Apex of
right lung

Apical bronchus
of right superior lobe

Right superior
lobar bronchus

Right middle
lobar bronchus

Right inferior
lobar bronchus

PLATE **42c** BRONCHOGRAM

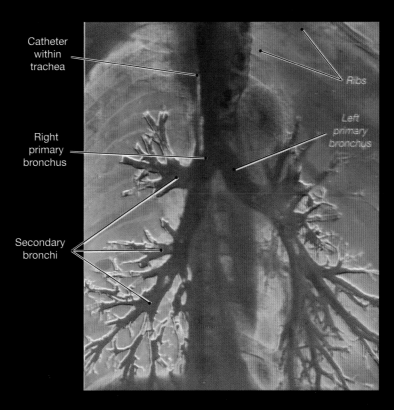

Catheter
within
trachea

Ribs

Left
primary
bronchus

Right
primary
bronchus

Secondary
bronchi

PLATE **42d** COLORIZED BRONCHOGRAM

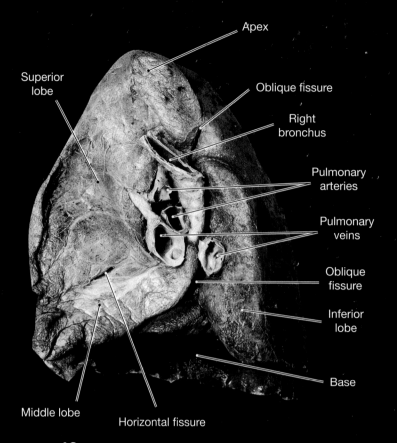

Apex

Superior
lobe

Oblique fissure

Right
bronchus

Pulmonary
arteries

Pulmonary
veins

Oblique
fissure

Inferior
lobe

Base

Middle lobe   Horizontal fissure

PLATE **43a** MEDIAL SURFACE OF RIGHT LUNG

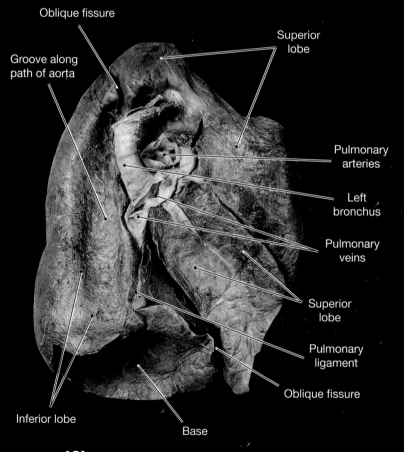

Oblique fissure

Superior
lobe

Groove along
path of aorta

Pulmonary
arteries

Left
bronchus

Pulmonary
veins

Superior
lobe

Pulmonary
ligament

Inferior lobe

Oblique fissure

Base

PLATE **43b** MEDIAL SURFACE OF LEFT LUNG

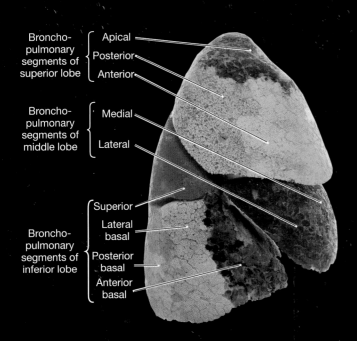

Broncho-
pulmonary
segments of
superior lobe — Apical
Posterior
Anterior

Broncho-
pulmonary
segments of
middle lobe — Medial
Lateral

Broncho-
pulmonary
segments of
inferior lobe — Superior
Lateral
basal
Posterior
basal
Anterior
basal

PLATE 43c BRONCHOPULMONARY SEGMENTS IN THE RIGHT LUNG, LATERAL VIEW

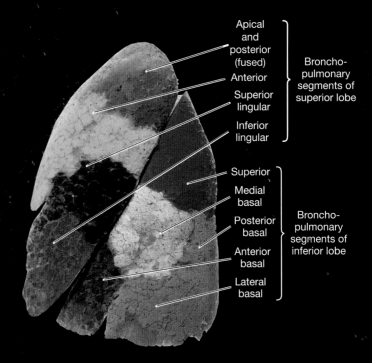

Apical
and
posterior
(fused)
Anterior
Superior
lingular
Inferior
lingular — Broncho-
pulmonary
segments of
superior lobe

Superior
Medial
basal
Posterior
basal
Anterior
basal
Lateral
basal — Broncho-
pulmonary
segments of
inferior lobe

PLATE 43d BRONCHOPULMONARY SEGMENTS IN THE LEFT LUNG, LATERAL VIEW

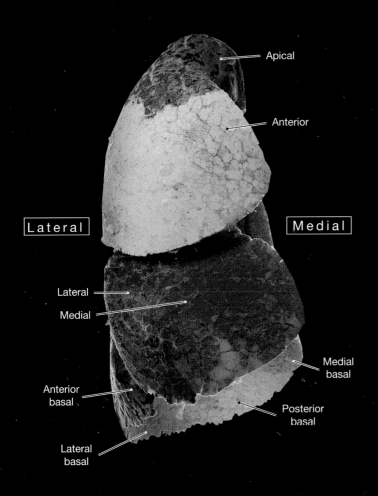

Apical

Anterior

Lateral

Medial

Lateral
Medial

Anterior
basal

Medial
basal

Posterior
basal

Lateral
basal

PLATE 44a BRONCHOPULMONARY SEGMENTS IN THE RIGHT LUNG, ANTERIOR VIEW

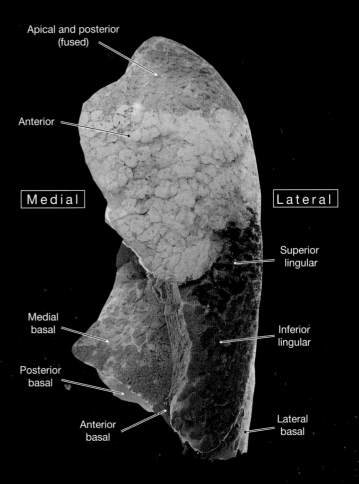

Apical and posterior
(fused)

Anterior

Medial

Lateral

Superior
lingular

Medial
basal

Inferior
lingular

Posterior
basal

Anterior
basal

Lateral
basal

PLATE 44b BRONCHOPULMONARY SEGMENTS IN THE LEFT LUNG, ANTERIOR VIEW

Catheter passing
through the
right atrium and
ventricle to enter
the pulmonary
trunk

Left
pulmonary
artery

Right
pulmonary
artery

Pulmonary
trunk

PLATE **44c** PULMONARY ANGIOGRAM

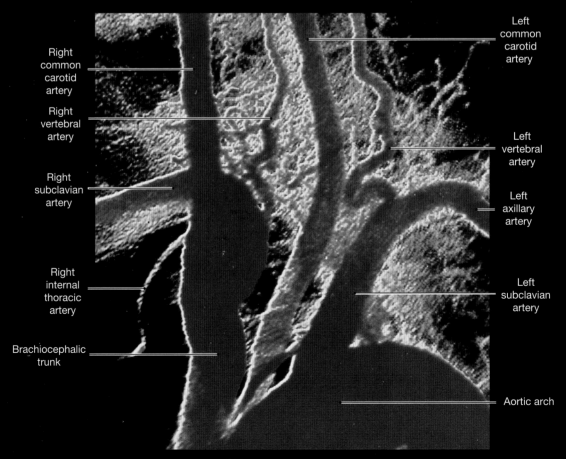

Right
common
carotid
artery

Right
vertebral
artery

Right
subclavian
artery

Right
internal
thoracic
artery

Brachiocephalic
trunk

Left
common
carotid
artery

Left
vertebral
artery

Left
axillary
artery

Left
subclavian
artery

Aortic arch

PLATE **45a** AORTIC ANGIOGRAM

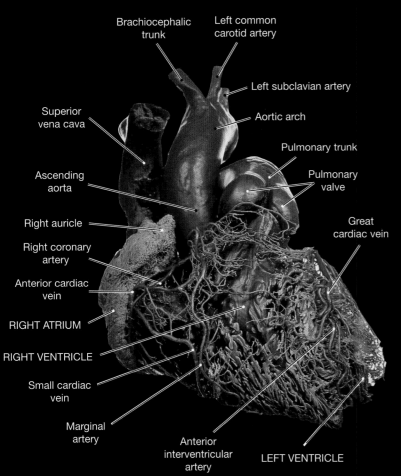

Brachiocephalic trunk

Left common carotid artery

Left subclavian artery

Aortic arch

Superior vena cava

Pulmonary trunk

Pulmonary valve

Ascending aorta

Right auricle

Great cardiac vein

Right coronary artery

Anterior cardiac vein

RIGHT ATRIUM

RIGHT VENTRICLE

Small cardiac vein

Marginal artery

Anterior interventricular artery

LEFT VENTRICLE

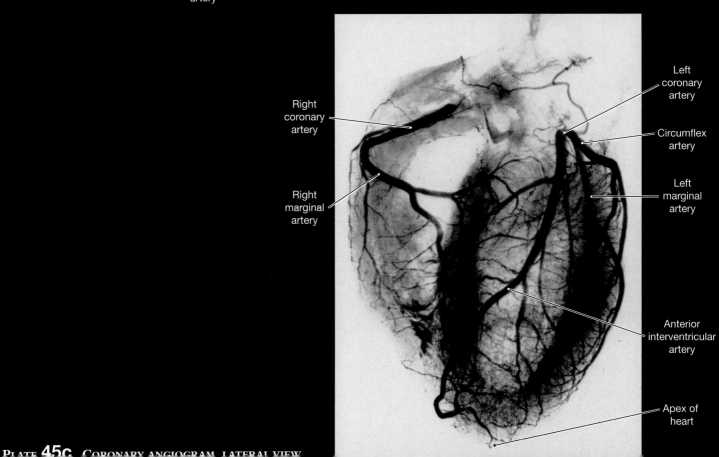

Right coronary artery

Left coronary artery

Circumflex artery

Right marginal artery

Left marginal artery

Anterior interventricular artery

Apex of heart

PLATE **45c** CORONARY ANGIOGRAM, LATERAL VIEW

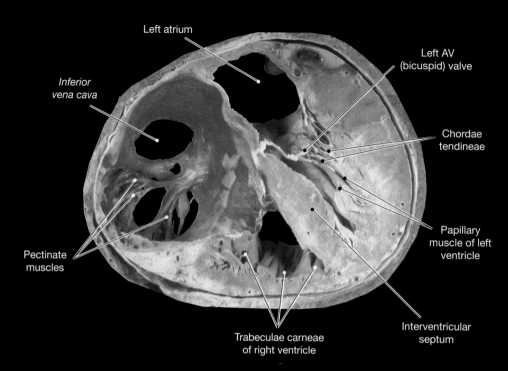

Left atrium

Inferior
vena cava

Left AV
(bicuspid) valve

Chordae
tendineae

Pectinate
muscles

Papillary
muscle of left
ventricle

Trabeculae carneae
of right ventricle

Interventricular
septum

PLATE **45d** HORIZONTAL SECTION THROUGH THE HEART, SUPERIOR VIEW

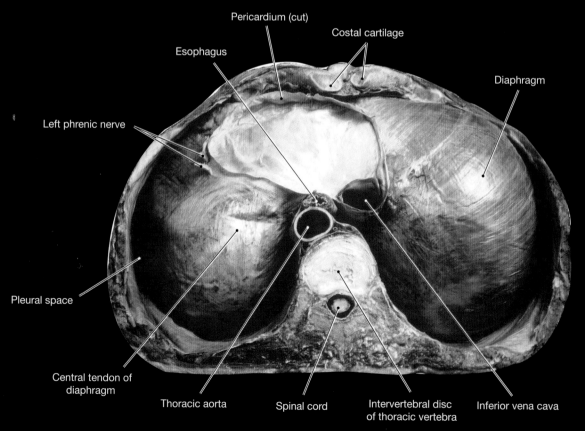

Pericardium (cut)

Costal cartilage

Esophagus

Diaphragm

Left phrenic nerve

Pleural space

Central tendon of
diaphragm

Thoracic aorta

Spinal cord

Intervertebral disc
of thoracic vertebra

Inferior vena cava

PLATE **46** DIAPHRAGM, SUPERIOR VIEW

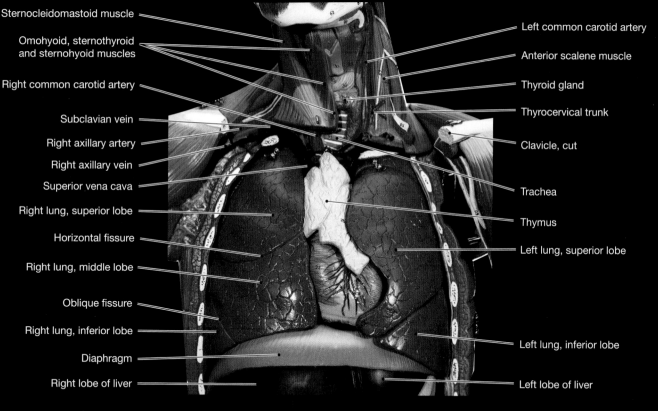

Sternocleidomastoid muscle

Omohyoid, sternothyroid and sternohyoid muscles

Right common carotid artery

Subclavian vein

Right axillary artery

Right axillary vein

Superior vena cava

Right lung, superior lobe

Horizontal fissure

Right lung, middle lobe

Oblique fissure

Right lung, inferior lobe

Diaphragm

Right lobe of liver

Left common carotid artery

Anterior scalene muscle

Thyroid gland

Thyrocervical trunk

Clavicle, cut

Trachea

Thymus

Left lung, superior lobe

Left lung, inferior lobe

Left lobe of liver

PLATE **47a** THORACIC ORGANS, SUPERFICIAL VIEW—MODEL

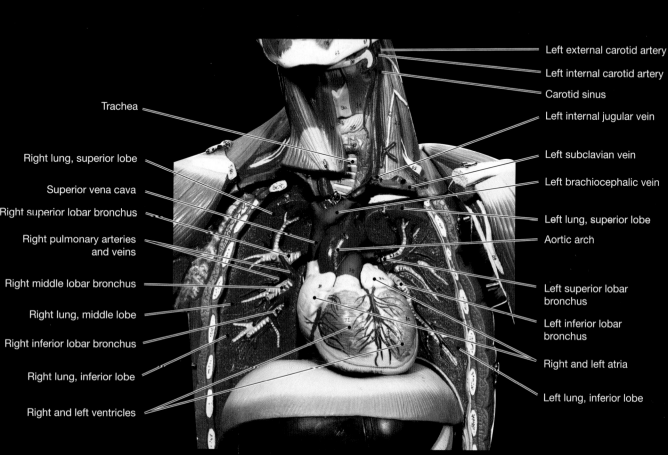

Trachea

Right lung, superior lobe

Superior vena cava

Right superior lobar bronchus

Right pulmonary arteries and veins

Right middle lobar bronchus

Right lung, middle lobe

Right inferior lobar bronchus

Right lung, inferior lobe

Right and left ventricles

Left external carotid artery

Left internal carotid artery

Carotid sinus

Left internal jugular vein

Left subclavian vein

Left brachiocephalic vein

Left lung, superior lobe

Aortic arch

Left superior lobar bronchus

Left inferior lobar bronchus

Right and left atria

Left lung, inferior lobe

PLATE **47b** THORACIC ORGANS, INTERMEDIATE VIEW—MODEL

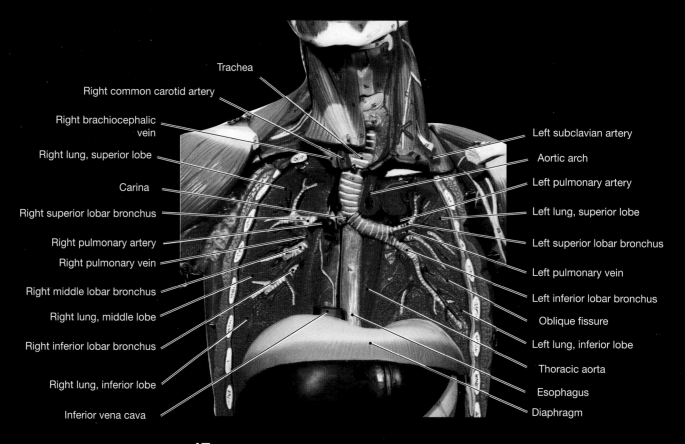

Trachea

Right common carotid artery

Right brachiocephalic vein

Right lung, superior lobe

Carina

Right superior lobar bronchus

Right pulmonary artery

Right pulmonary vein

Right middle lobar bronchus

Right lung, middle lobe

Right inferior lobar bronchus

Right lung, inferior lobe

Inferior vena cava

Left subclavian artery

Aortic arch

Left pulmonary artery

Left lung, superior lobe

Left superior lobar bronchus

Left pulmonary vein

Left inferior lobar bronchus

Oblique fissure

Left lung, inferior lobe

Thoracic aorta

Esophagus

Diaphragm

PLATE 47c THORACIC ORGANS, DEEPER VIEW—MODEL

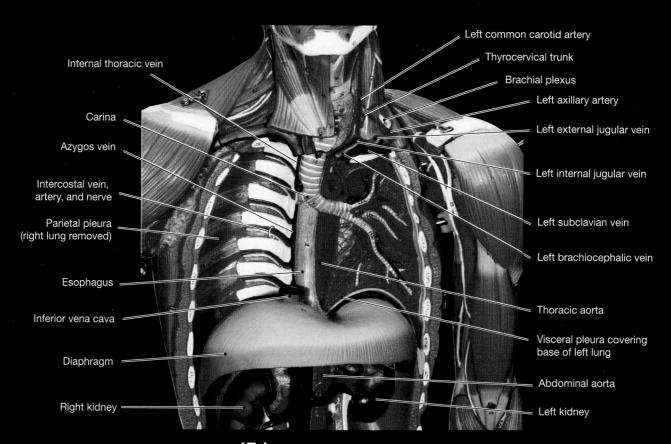

Internal thoracic vein

Carina

Azygos vein

Intercostal vein, artery, and nerve

Parietal pleura (right lung removed)

Esophagus

Inferior vena cava

Diaphragm

Right kidney

Left common carotid artery

Thyrocervical trunk

Brachial plexus

Left axillary artery

Left external jugular vein

Left internal jugular vein

Left subclavian vein

Left brachiocephalic vein

Thoracic aorta

Visceral pleura covering base of left lung

Abdominal aorta

Left kidney

PLATE 47d THE THORACIC CAVITY

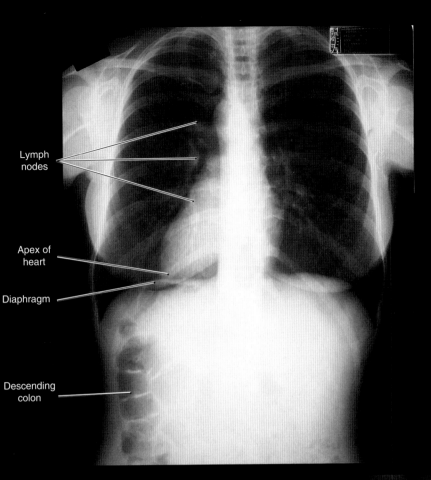

Lymph
nodes

Apex of
heart

Diaphragm

Descending
colon

PLATE **48b** LYMPHANGIOGRAM OF THORAX,
ANTERIOR–POSTERIOR PROJECTION

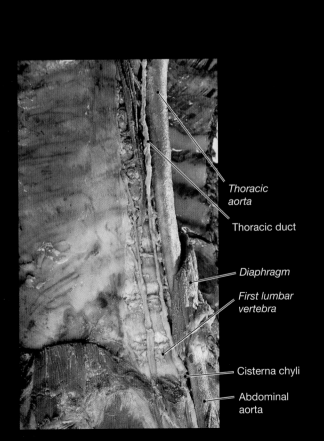

*Thoracic
aorta*

Thoracic duct

Diaphragm

*First lumbar
vertebra*

Cisterna chyli

Abdominal
aorta

Plate 48c provided by NetAnatomy.com.

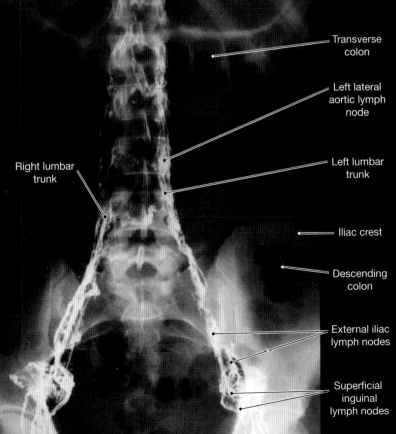

Transverse
colon

Left lateral
aortic lymph
node

Left lumbar
trunk

Right lumbar
trunk

Iliac crest

Descending
colon

External iliac
lymph nodes

Superficial
inguinal
lymph nodes

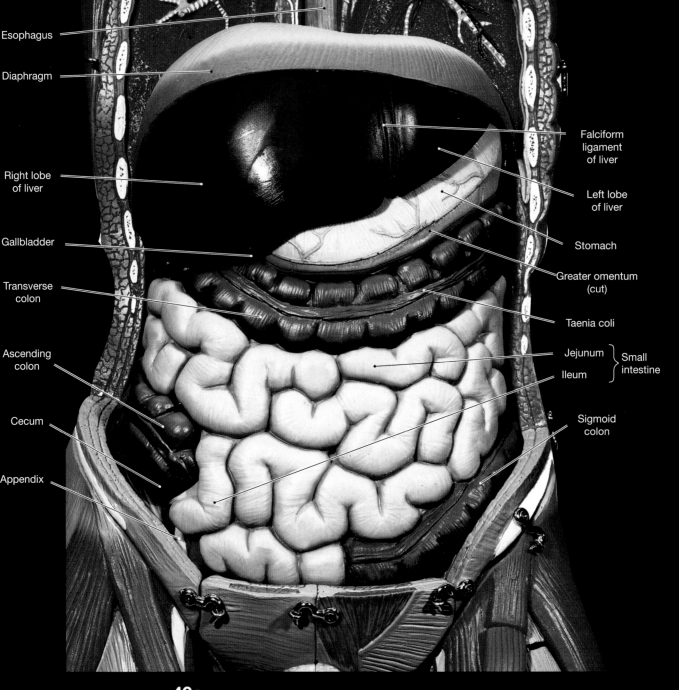

Esophagus

Diaphragm

Right lobe
of liver

Gallbladder

Transverse
colon

Ascending
colon

Cecum

Appendix

Falciform
ligament
of liver

Left lobe
of liver

Stomach

Greater omentum
(cut)

Taenia coli

Jejunum

Ileum

} Small
intestine

Sigmoid
colon

PLATE **49a** THE ABDOMINOPELVIC VISCERA, SUPERFICIAL ANTERIOR VIEW

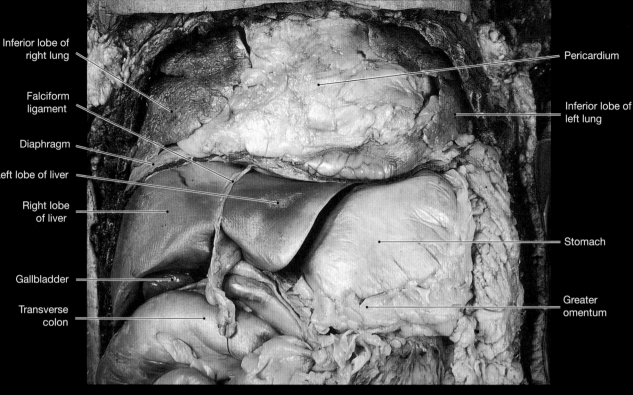

Inferior lobe of right lung

Falciform ligament

Diaphragm

Left lobe of liver

Right lobe of liver

Gallbladder

Transverse colon

Pericardium

Inferior lobe of left lung

Stomach

Greater omentum

PLATE 49b SUPERIOR PORTION OF ABDOMINOPELVIC CAVITY, ANTERIOR VIEW

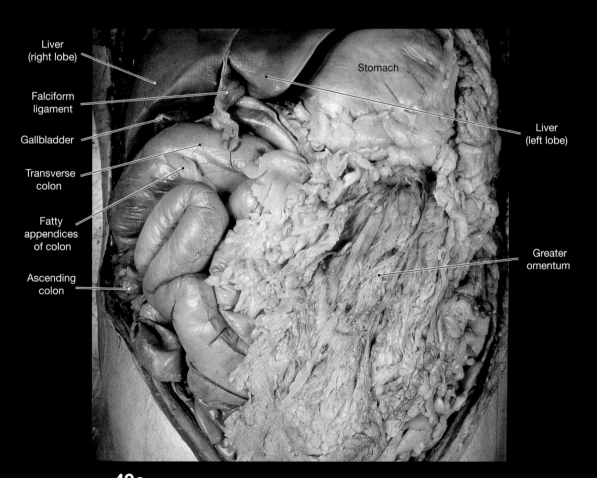

Liver (right lobe)

Falciform ligament

Gallbladder

Transverse colon

Fatty appendices of colon

Ascending colon

Stomach

Liver (left lobe)

Greater omentum

PLATE 49c INFERIOR PORTION OF ABDOMINOPELVIC CAVITY, ANTERIOR VIEW

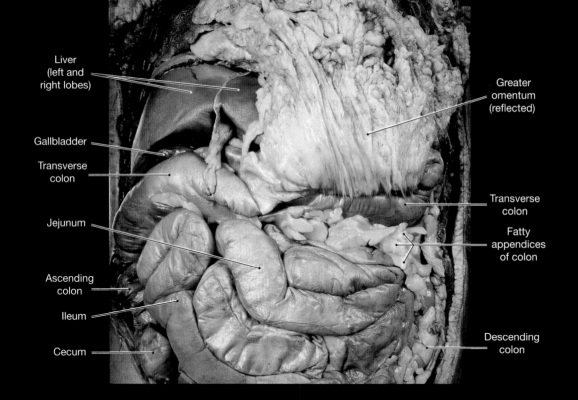

Liver
(left and
right lobes)

Gallbladder

Transverse
colon

Jejunum

Ascending
colon

Ileum

Cecum

Greater
omentum
(reflected)

Transverse
colon

Fatty
appendices
of colon

Descending
colon

PLATE 49d ABDOMINAL DISSECTION, GREATER OMENTUM REFLECTED

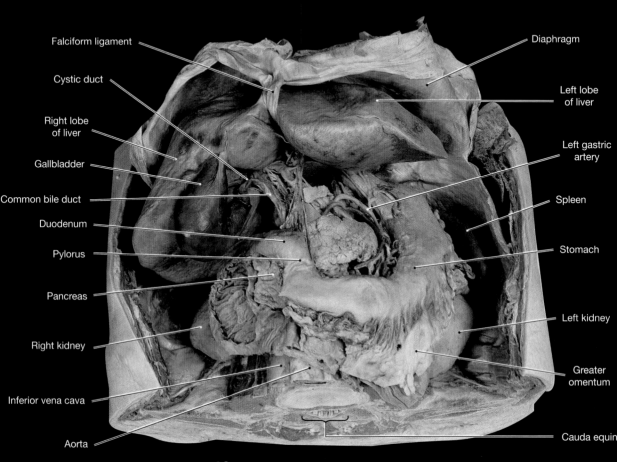

Falciform ligament

Cystic duct

Right lobe
of liver

Gallbladder

Common bile duct

Duodenum

Pylorus

Pancreas

Right kidney

Inferior vena cava

Aorta

Diaphragm

Left lobe
of liver

Left gastric
artery

Spleen

Stomach

Left kidney

Greater
omentum

Cauda equin

PLATE 49e LIVER AND GALLBLADDER IN SITU

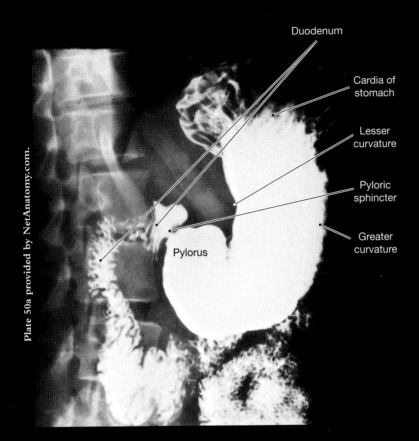

Duodenum

Cardia of
stomach

Lesser
curvature

Pyloric
sphincter

Greater
curvature

Pylorus

PLATE **50a** GASTRIC RADIOGRAM 1

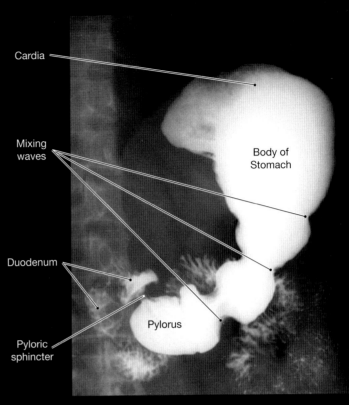

Cardia

Mixing
waves

Body of
Stomach

Duodenum

Pylorus

Pyloric
sphincter

PLATE **50b** GASTRIC RADIOGRAM 2

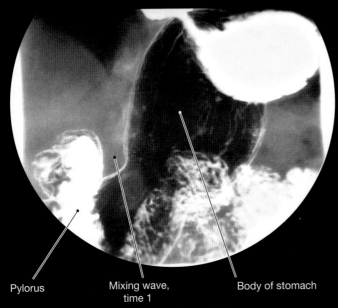

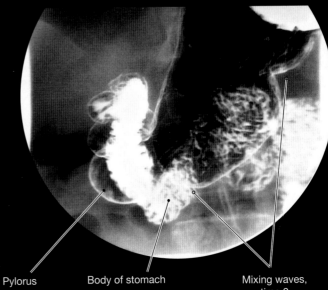

Pylorus

Mixing wave,
time 1

Body of stomach

Pylorus

Body of stomach

Mixing waves,
time 2

PLATE **50c** GASTRIC MOTILITY ANTERIOR VIEW TIMES 1 AND 2

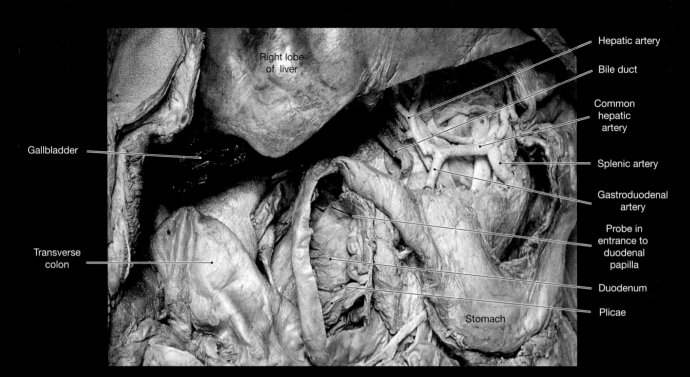

Right lobe
of liver

Hepatic artery

Bile duct

Common
hepatic
artery

Splenic artery

Gastroduodenal
artery

Probe in
entrance to
duodenal
papilla

Duodenum

Plicae

Gallbladder

Transverse
colon

Stomach

PLATE 51a ABDOMINAL DISSECTION, DUODENAL REGION

PLATE 51b GROSS ANATOMY
OF DUODENAL MUCOSA

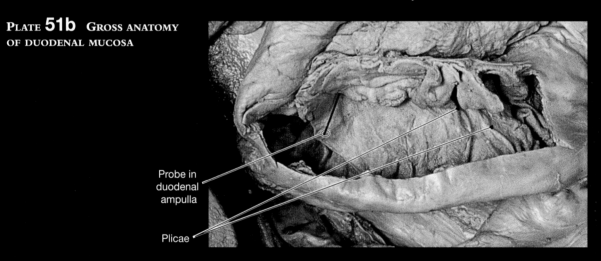

Probe in
duodenal
ampulla

Plicae

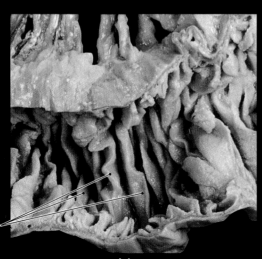

Plicae

Jejunum

PLATE 51c GROSS ANATOMY OF THE JEJUNAL MUCOSA

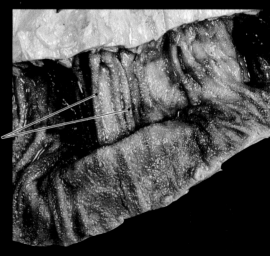

Plicae

Ileum

PLATE 51d GROSS ANATOMY OF THE ILEAL MUCOSA

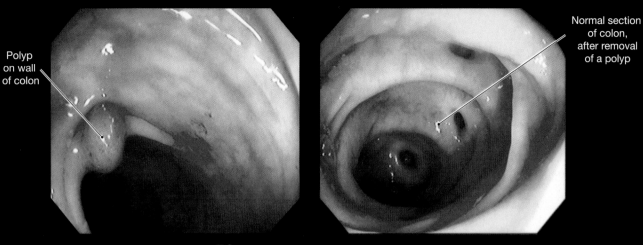

Polyp on wall of colon

Normal section of colon, after removal of a polyp

PLATE 52 NORMAL AND ABNORMAL COLONOSCOPE

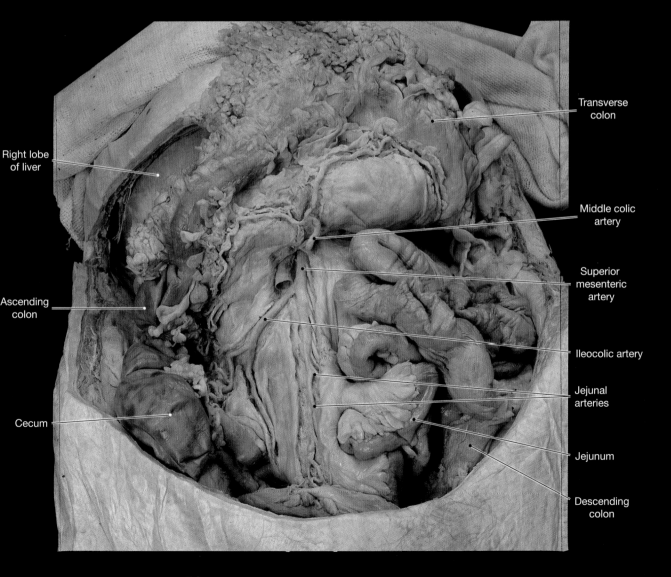

Transverse colon

Right lobe of liver

Middle colic artery

Superior mesenteric artery

Ascending colon

Ileocolic artery

Jejunal arteries

Cecum

Jejunum

Descending colon

PLATE 53a BRANCHES OF THE SUPERIOR MESENTERIC ARTERY

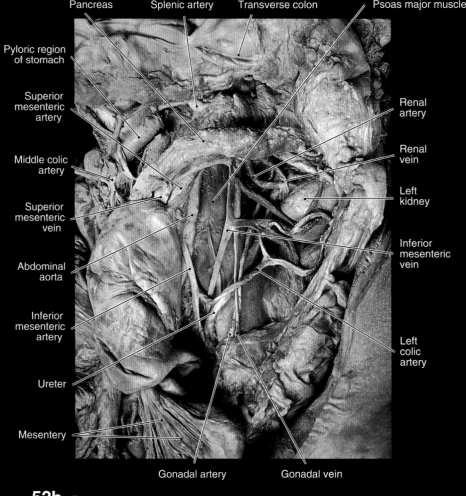

Pancreas   Splenic artery   Transverse colon   Psoas major muscle

Pyloric region
of stomach

Superior
mesenteric
artery

Middle colic
artery

Superior
mesenteric
vein

Abdominal
aorta

Inferior
mesenteric
artery

Ureter

Mesentery

Renal
artery

Renal
vein

Left
kidney

Inferior
mesenteric
vein

Left
colic
artery

Gonadal artery   Gonadal vein

PLATE 53b   INFERIOR MESENTERIC VESSELS

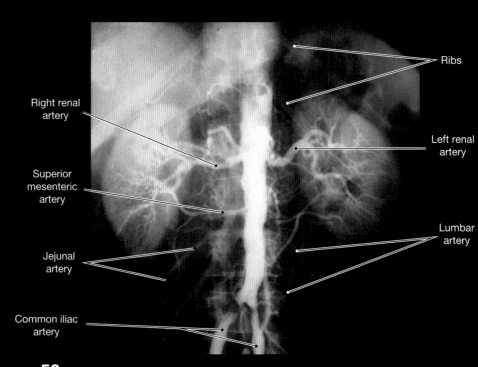

Ribs

Right renal
artery

Left renal
artery

Superior
mesenteric
artery

Jejunal
artery

Lumbar
artery

Common iliac
artery

PLATE 53c   ABDOMINAL ARTERIOGRAM 1

PLATE **53d** ABDOMINAL
ARTERIOGRAM 2

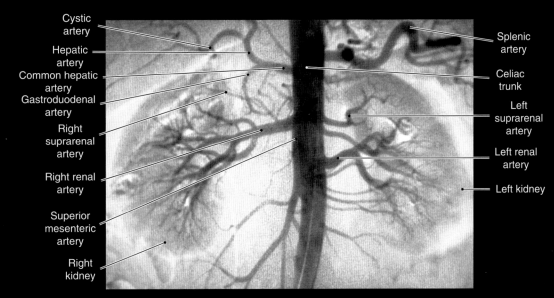

Cystic
artery

Hepatic
artery

Common hepatic
artery

Gastroduodenal
artery

Right
suprarenal
artery

Right renal
artery

Superior
mesenteric
artery

Right
kidney

Splenic
artery

Celiac
trunk

Left
suprarenal
artery

Left renal
artery

Left kidney

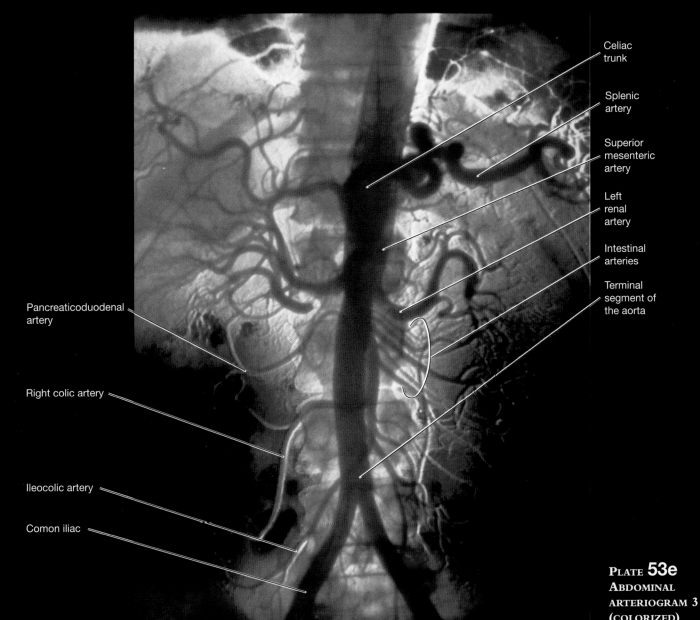

Celiac
trunk

Splenic
artery

Superior
mesenteric
artery

Left
renal
artery

Intestinal
arteries

Terminal
segment of
the aorta

Pancreaticoduodenal
artery

Right colic artery

Ileocolic artery

Comon iliac

PLATE **53e**
ABDOMINAL
ARTERIOGRAM 3
(COLORIZED)

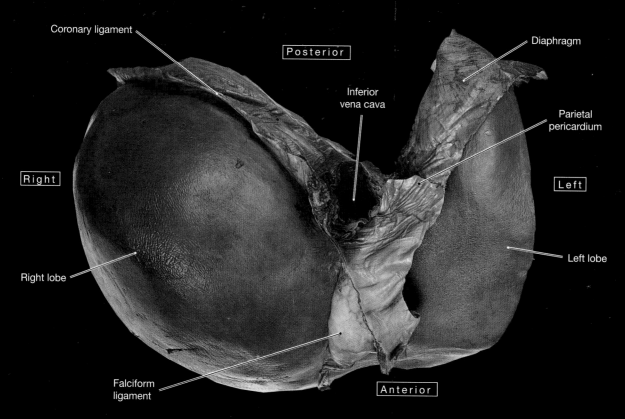

Coronary ligament

Posterior

Inferior
vena cava

Diaphragm

Parietal
pericardium

Right

Left

Left lobe

Right lobe

Falciform
ligament

Anterior

PLATE 54a THE ISOLATED LIVER AND GALLBLADDER, SUPERIOR VIEW

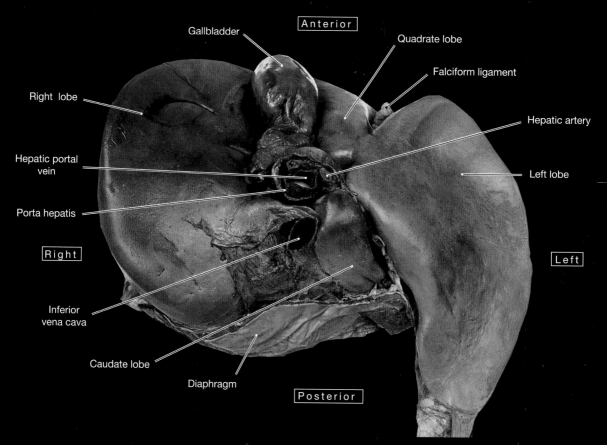

Anterior

Gallbladder

Quadrate lobe

Falciform ligament

Right lobe

Hepatic artery

Hepatic portal
vein

Left lobe

Porta hepatis

Right

Left

Inferior
vena cava

Caudate lobe

Diaphragm

Posterior

PLATE 54b THE ISOLATED LIVER AND GALLBLADDER, INFERIOR VIEW

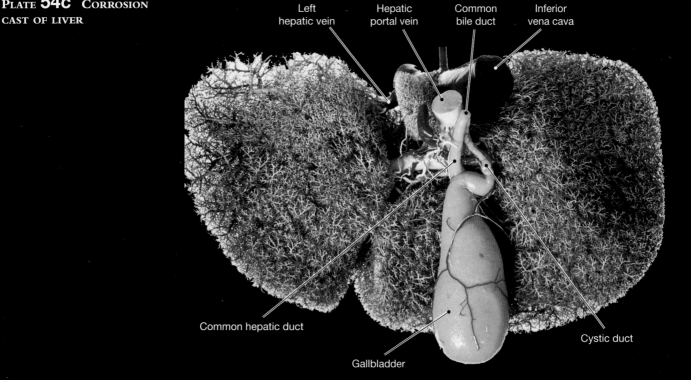

Left
hepatic vein

Hepatic
portal vein

Common
bile duct

Inferior
vena cava

Common hepatic duct

Gallbladder

Cystic duct

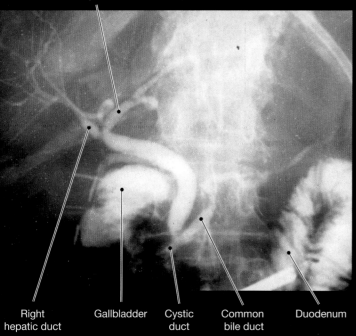

Left hepatic duct

Right
hepatic duct

Gallbladder

Cystic
duct

Common
bile duct

Duodenum

PLATE **54d** CHOLANGIOPAN-CREATOGRAM (PANCREATIC
AND BILE DUCTS)

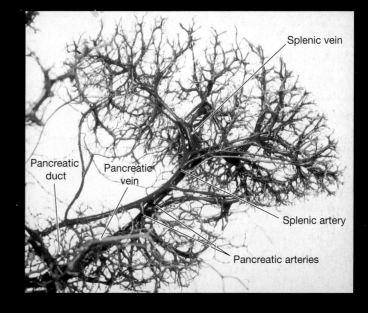

Splenic vein

Pancreatic
duct

Pancreatic
vein

Splenic artery

Pancreatic arteries

PLATE **55a** CORROSION CAST OF SPLENIC AND PANCREATIC
VESSELS

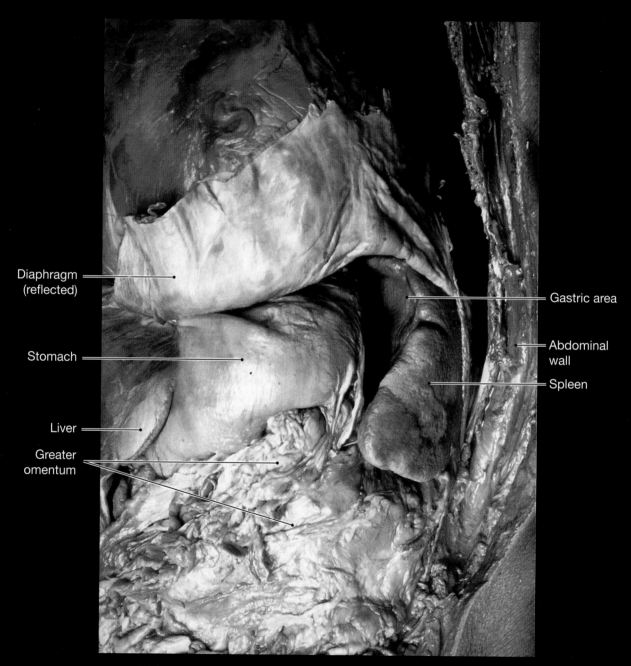

Diaphragm (reflected)

Stomach

Liver

Greater omentum

Gastric area

Abdominal wall

Spleen

PLATE **55b** SPLEEN, ANTERIOR VIEW

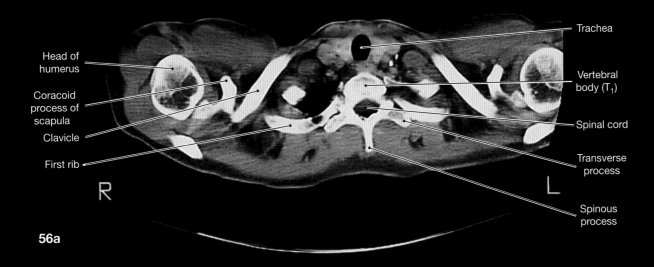

Trachea

Head of humerus

Coracoid process of scapula

Clavicle

First rib

Vertebral body (T_1)

Spinal cord

Transverse process

Spinous process

R

L

56a

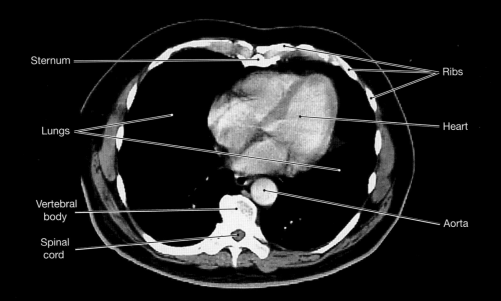

Sternum

Ribs

Lungs

Heart

Vertebral body

Aorta

Spinal cord

56b

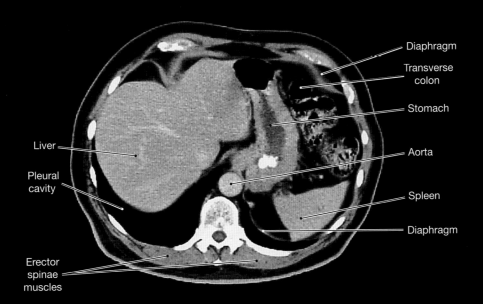

Diaphragm

Transverse colon

Stomach

Liver

Aorta

Pleural cavity

Spleen

Diaphragm

Erector spinae muscles

56c

PLATES 56a–c MRI SCANS OF THE TRUNK, HORIZONTAL SECTIONS, SUPERIOR TO INFERIOR SEQUENCE

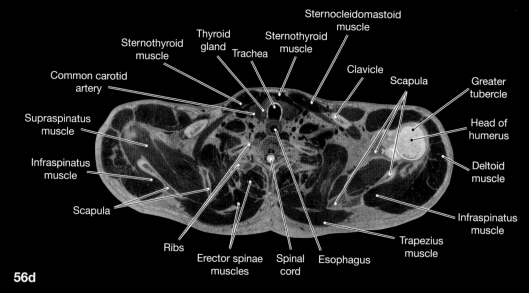

Sternocleidomastoid muscle

Thyroid gland

Sternothyroid muscle

Sternothyroid muscle

Trachea

Clavicle

Scapula

Greater tubercle

Common carotid artery

Head of humerus

Supraspinatus muscle

Deltoid muscle

Infraspinatus muscle

Infraspinatus muscle

Scapula

Trapezius muscle

Ribs

Erector spinae muscles

Spinal cord

Esophagus

56d

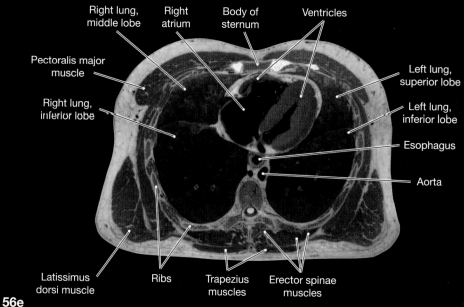

Right lung, middle lobe

Right atrium

Body of sternum

Ventricles

Pectoralis major muscle

Left lung, superior lobe

Right lung, inferior lobe

Left lung, inferior lobe

Esophagus

Aorta

Latissimus dorsi muscle

Ribs

Trapezius muscles

Erector spinae muscles

56e

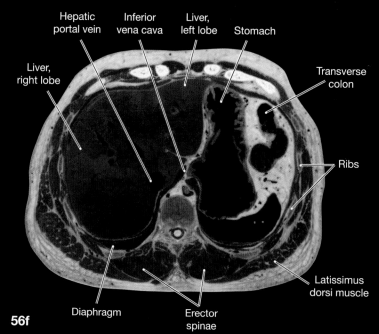

Hepatic portal vein

Inferior vena cava

Liver, left lobe

Stomach

Liver, right lobe

Transverse colon

Ribs

Latissimus dorsi muscle

Diaphragm

Erector spinae

56f

PLATE **56d–f** HORIZONTAL SECTIONS THROUGH THE TRUNK.

These sections, derived from the Visible Human dataset, are at the approximate levels of the scans shown in parts a–c.

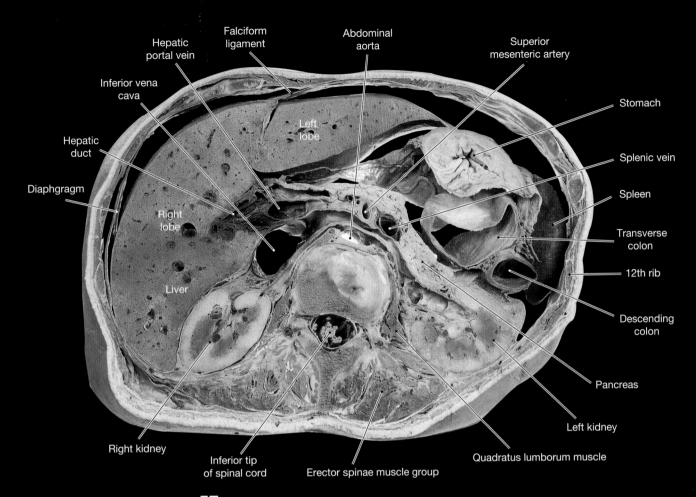

Hepatic portal vein

Falciform ligament

Abdominal aorta

Superior mesenteric artery

Inferior vena cava

Stomach

Splenic vein

Hepatic duct

Spleen

Diaphgragm

Left lobe

Right lobe

Transverse colon

12th rib

Liver

Descending colon

Pancreas

Left kidney

Right kidney

Inferior tip of spinal cord

Erector spinae muscle group

Quadratus lumborum muscle

PLATE 57a ABDOMINAL CAVITY, HORIZONTAL SECTION AT T$_{12}$

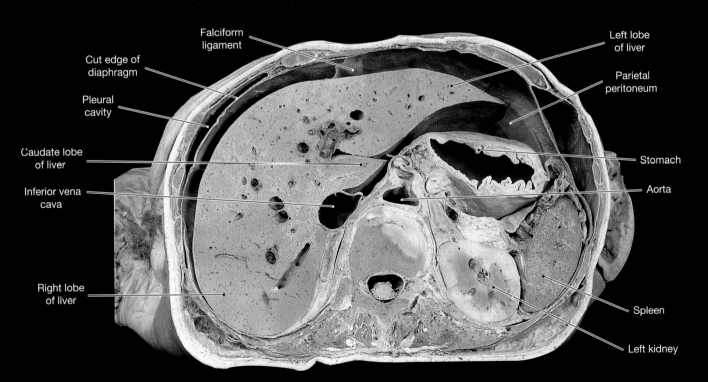

Falciform ligament

Left lobe of liver

Cut edge of diaphragm

Parietal peritoneum

Pleural cavity

Caudate lobe of liver

Stomach

Inferior vena cava

Aorta

Right lobe of liver

Spleen

Left kidney

PLATE 57b ABDOMINAL CAVITY, HORIZONTAL SECTION AT L$_1$

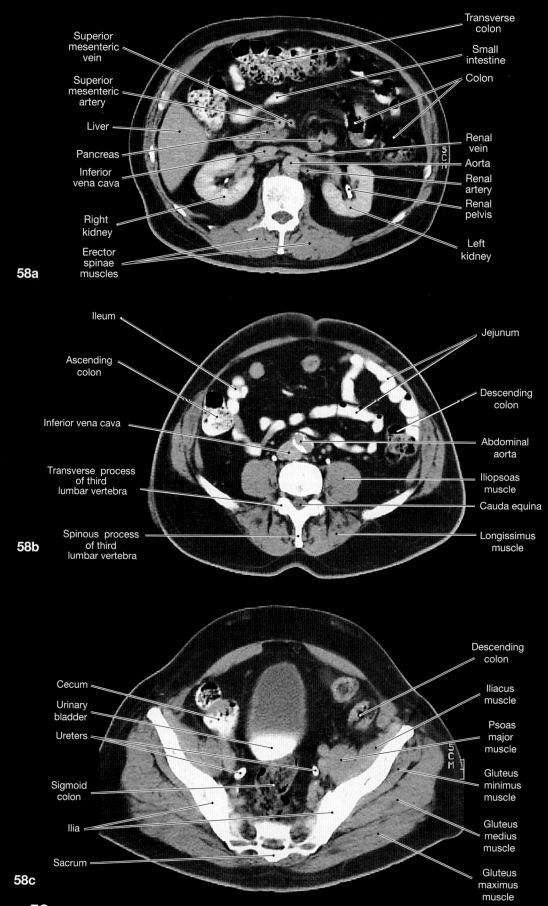

Transverse colon

Superior mesenteric vein

Superior mesenteric artery

Small intestine

Colon

Liver

Pancreas

Renal vein

Aorta

Inferior vena cava

Renal artery

Renal pelvis

Right kidney

Left kidney

Erector spinae muscles

58a

Ileum

Jejunum

Ascending colon

Descending colon

Inferior vena cava

Abdominal aorta

Transverse process of third lumbar vertebra

Iliopsoas muscle

Cauda equina

Spinous process of third lumbar vertebra

Longissimus muscle

58b

Descending colon

Cecum

Iliacus muscle

Urinary bladder

Psoas major muscle

Ureters

Gluteus minimus muscle

Sigmoid colon

Ilia

Gluteus medius muscle

Sacrum

Gluteus maximus muscle

58c

PLATES **58a–c**   MRI SCANS OF THE TRUNK, HORIZONTAL SECTIONS, SUPERIOR TO INFERIOR SEQUENCE

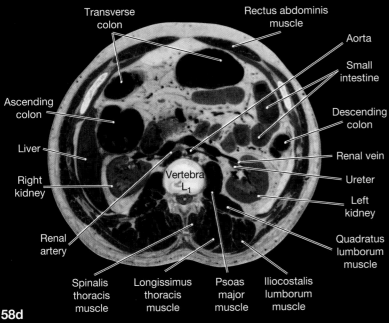

Transverse colon
Rectus abdominis muscle
Aorta
Small intestine
Ascending colon
Descending colon
Liver
Renal vein
Right kidney
Ureter
Left kidney
Renal artery
Quadratus lumborum muscle
Vertebra L₁
Spinalis thoracis muscle
Longissimus thoracis muscle
Psoas major muscle
Iliocostalis lumborum muscle

58d

PLATE 58d–f HORIZONTAL SECTIONS THROUGH THE TRUNK.

These sections, derived from the Visible Human dataset, are at the approximate levels of the scans shown in part a–c.

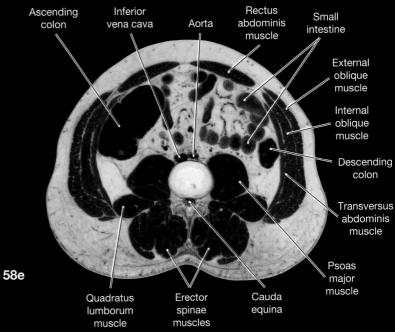

Ascending colon
Inferior vena cava
Aorta
Rectus abdominis muscle
Small intestine
External oblique muscle
Internal oblique muscle
Descending colon
Transversus abdominis muscle
Psoas major muscle
Quadratus lumborum muscle
Erector spinae muscles
Cauda equina

58e

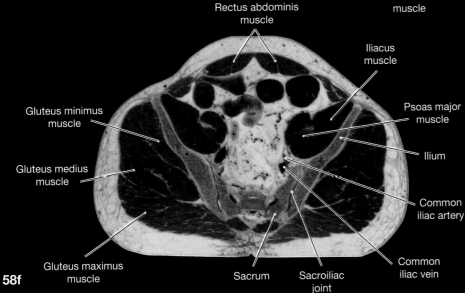

Rectus abdominis muscle
Iliacus muscle
Gluteus minimus muscle
Psoas major muscle
Gluteus medius muscle
Ilium
Common iliac artery
Gluteus maximus muscle
Sacrum
Sacroiliac joint
Common iliac vein

58f

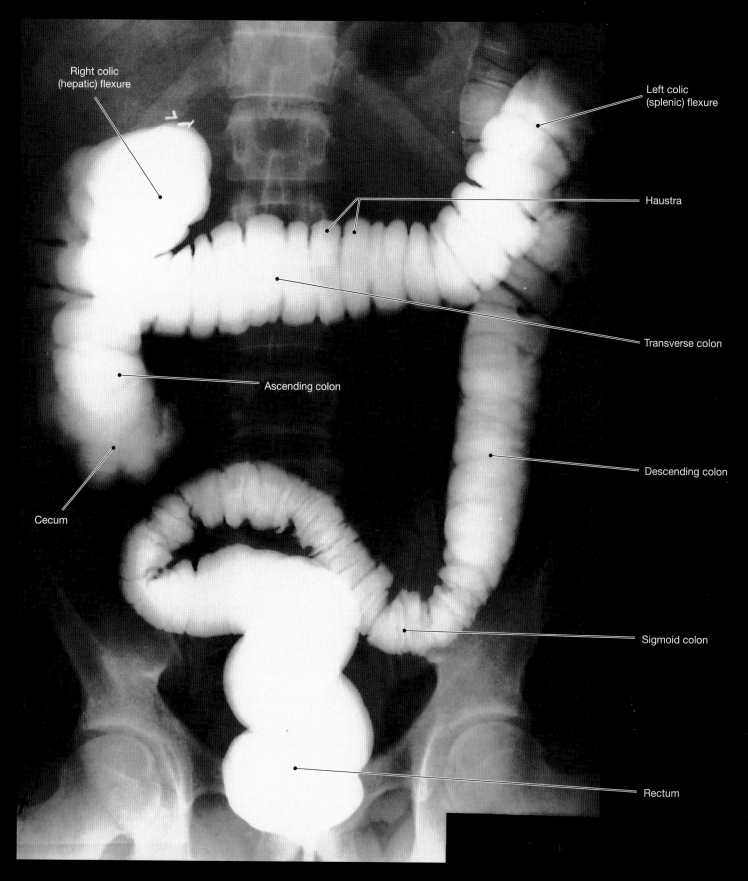

Right colic
(hepatic) flexure

Left colic
(splenic) flexure

Haustra

Transverse colon

Ascending colon

Descending colon

Cecum

Sigmoid colon

Rectum

PLATE 59   CONTRAST X-RAY OF COLON AND RECTUM, ANTERIOR-POSTERIOR PROJECTION

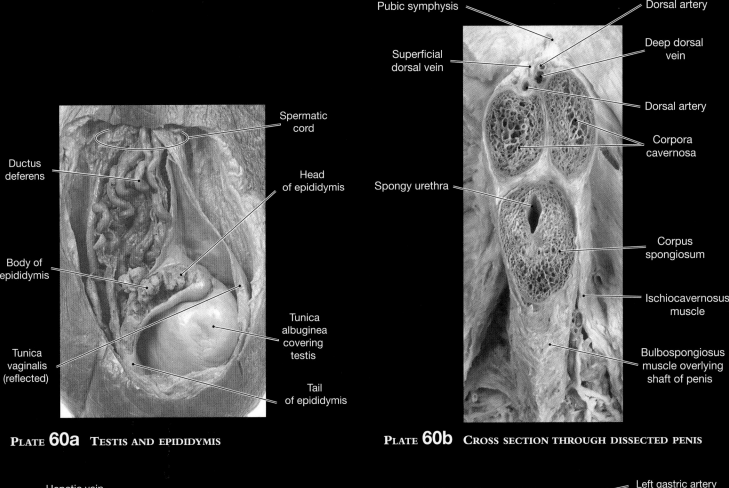

Plate 60a Testis and epididymis

Spermatic cord

Ductus deferens

Head of epididymis

Body of epididymis

Tunica albuginea covering testis

Tunica vaginalis (reflected)

Tail of epididymis

Plate 60b Cross section through dissected penis

Pubic symphysis

Dorsal artery

Superficial dorsal vein

Deep dorsal vein

Dorsal artery

Corpora cavernosa

Spongy urethra

Corpus spongiosum

Ischiocavernosus muscle

Bulbospongiosus muscle overlying shaft of penis

Plate 61a The posterior abdominal wall

Hepatic vein (stump)

Left renal vein

Right adrenal gland

Inferior vena cava

Right renal vein

Right renal artery

Right kidney

Peritoneum

Right ureter

Right gonadal vein

Inferior mesenteric artery

Left gastric artery

Common hepatic artery

Splenic artery

Celiac trunk

Celiac ganglion

Left adrenal gland

Left suprarenal vein

Left renal vein

Left renal artery

Left kidney

Superior mesenteric artery

Left ureter

Left gonadal vein

Gonadal arteries

Abdominal aorta

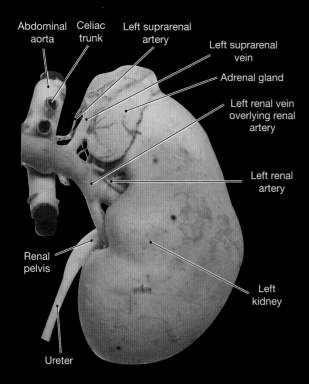

Abdominal aorta

Celiac trunk

Left suprarenal artery

Left suprarenal vein

Adrenal gland

Left renal vein overlying renal artery

Left renal artery

Renal pelvis

Left kidney

Ureter

PLATE 61b LEFT KIDNEY, ANTERIOR VIEW

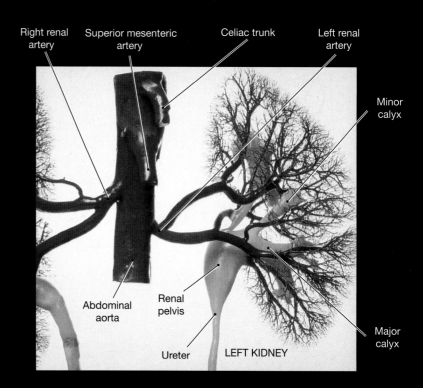

Right renal artery

Superior mesenteric artery

Celiac trunk

Left renal artery

Minor calyx

Abdominal aorta

Renal pelvis

Ureter

LEFT KIDNEY

Major calyx

PLATE 61c CORROSION CAST OF THE RENAL ARTERIES, URETERS, AND RENAL PELVIS

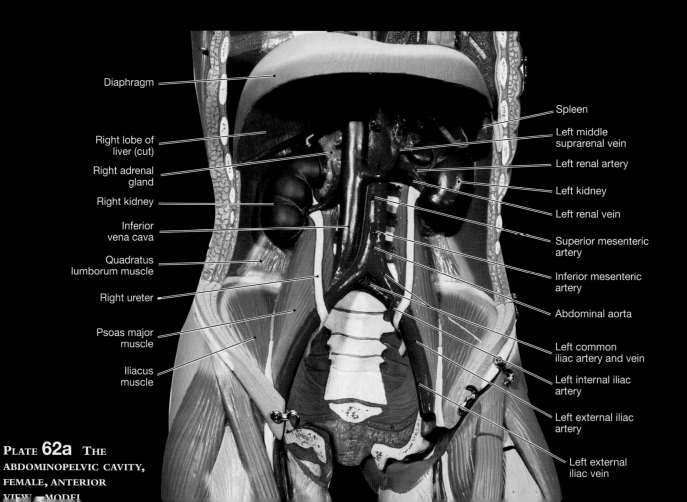

Diaphragm

Right lobe of liver (cut)

Right adrenal gland

Right kidney

Inferior vena cava

Quadratus lumborum muscle

Right ureter

Psoas major muscle

Iliacus muscle

Spleen

Left middle suprarenal vein

Left renal artery

Left kidney

Left renal vein

Superior mesenteric artery

Inferior mesenteric artery

Abdominal aorta

Left common iliac artery and vein

Left internal iliac artery

Left external iliac artery

Left external iliac vein

PLATE 62a THE ABDOMINOPELVIC CAVITY, FEMALE, ANTERIOR VIEW—MODEL

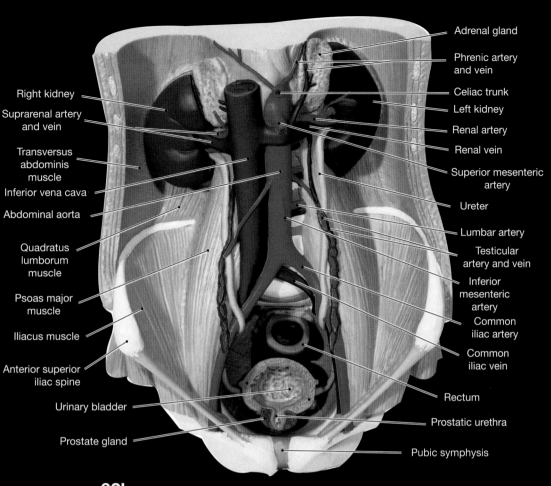

Adrenal gland

Phrenic artery
and vein

Celiac trunk

Left kidney

Renal artery

Renal vein

Superior mesenteric
artery

Ureter

Lumbar artery

Testicular
artery and vein

Inferior
mesenteric
artery

Common
iliac artery

Common
iliac vein

Rectum

Prostatic urethra

Pubic symphysis

Right kidney

Suprarenal artery
and vein

Transversus
abdominis
muscle

Inferior vena cava

Abdominal aorta

Quadratus
lumborum
muscle

Psoas major
muscle

Iliacus muscle

Anterior superior
iliac spine

Urinary bladder

Prostate gland

PLATE **62b** THE ABDOMINOPELVIC CAVITY, MALE, ANTERIOR VIEW—MODEL

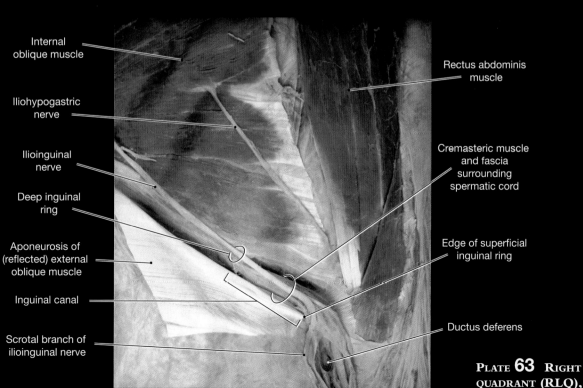

Internal
oblique muscle

Iliohypogastric
nerve

Ilioinguinal
nerve

Deep inguinal
ring

Aponeurosis of
(reflected) external
oblique muscle

Inguinal canal

Scrotal branch of
ilioinguinal nerve

Rectus abdominis
muscle

Cremasteric muscle
and fascia
surrounding
spermatic cord

Edge of superficial
inguinal ring

Ductus deferens

PLATE **63** RIGHT
QUADRANT (RLQ),

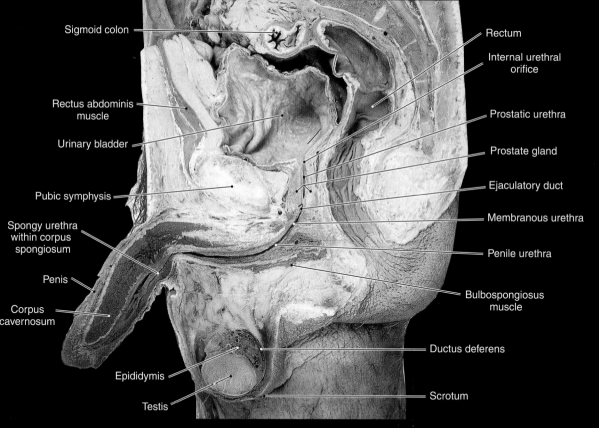

Sigmoid colon

Rectus abdominis muscle

Urinary bladder

Pubic symphysis

Spongy urethra within corpus spongiosum

Penis

Corpus cavernosum

Epididymis

Testis

Rectum

Internal urethral orifice

Prostatic urethra

Prostate gland

Ejaculatory duct

Membranous urethra

Penile urethra

Bulbospongiosus muscle

Ductus deferens

Scrotum

PLATE 64 PELVIC REGION OF A MALE, SAGITTAL SECTION

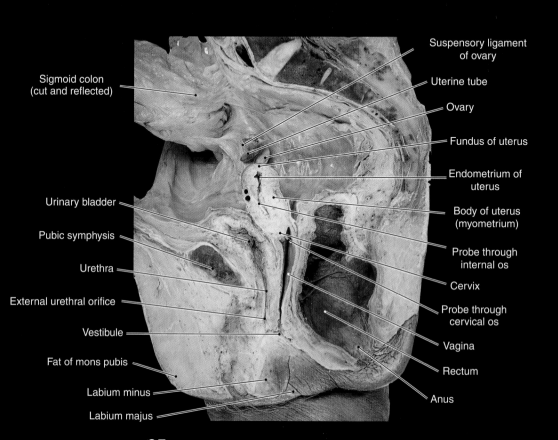

Sigmoid colon (cut and reflected)

Urinary bladder

Pubic symphysis

Urethra

External urethral orifice

Vestibule

Fat of mons pubis

Labium minus

Labium majus

Suspensory ligament of ovary

Uterine tube

Ovary

Fundus of uterus

Endometrium of uterus

Body of uterus (myometrium)

Probe through internal os

Cervix

Probe through cervical os

Vagina

Rectum

Anus

PLATE 65 PELVIC REGION OF A FEMALE, SAGITTAL SECTION

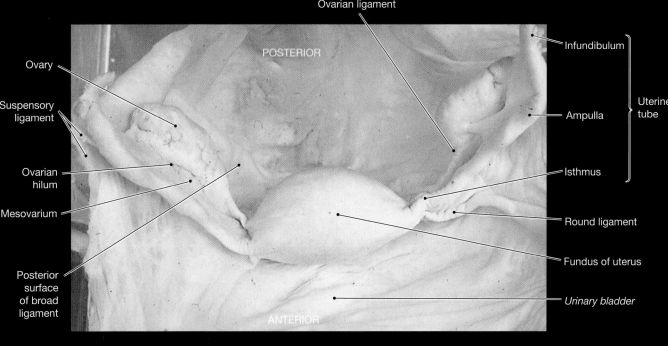

Ovarian ligament

Ovary

Suspensory ligament

Ovarian hilum

Mesovarium

Posterior surface of broad ligament

POSTERIOR

Infundibulum

Ampulla

Uterine tube

Isthmus

Round ligament

Fundus of uterus

Urinary bladder

ANTERIOR

PLATE 66 PELVIC CAVITY, SUPERIOR VIEW

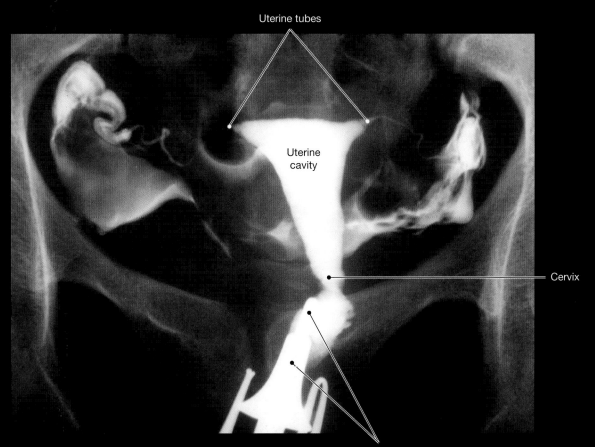

Uterine tubes

Uterine cavity

Cervix

Application tube in vagina, source of contrast medium

PLATE 67 HYSTEROSALPINGOGRAM

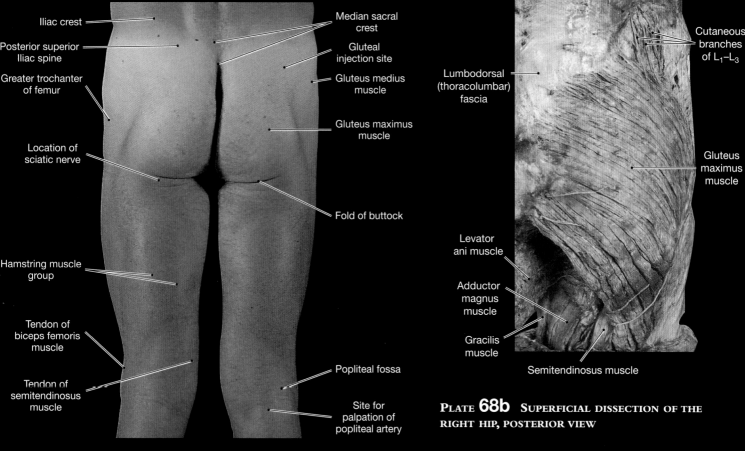

Iliac crest

Posterior superior Iliac spine

Greater trochanter of femur

Location of sciatic nerve

Hamstring muscle group

Tendon of biceps femoris muscle

Tendon of semitendinosus muscle

Median sacral crest

Gluteal injection site

Gluteus medius muscle

Gluteus maximus muscle

Fold of buttock

Popliteal fossa

Site for palpation of popliteal artery

Lumbodorsal (thoracolumbar) fascia

Cutaneous branches of L_1–L_3

Gluteus maximus muscle

Levator ani muscle

Adductor magnus muscle

Gracilis muscle

Semitendinosus muscle

PLATE 68b SUPERFICIAL DISSECTION OF THE RIGHT HIP, POSTERIOR VIEW

PLATE 68a GLUTEAL REGION AND THIGHS

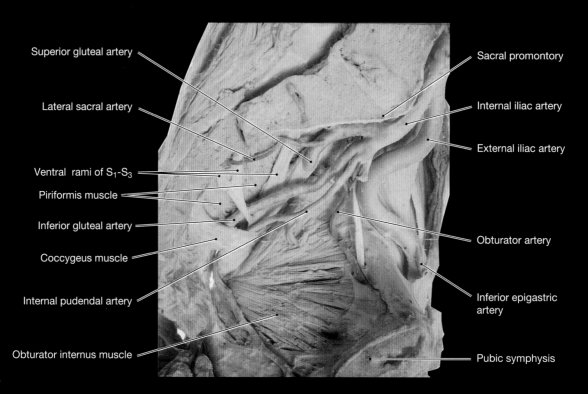

Superior gluteal artery

Lateral sacral artery

Ventral rami of S_1-S_3

Piriformis muscle

Inferior gluteal artery

Coccygeus muscle

Internal pudendal artery

Obturator internus muscle

Sacral promontory

Internal iliac artery

External iliac artery

Obturator artery

Inferior epigastric artery

Pubic symphysis

PLATE 68c BLOOD VESSELS, NERVES, AND MUSCLES IN THE LEFT HALF OF THE PELVIS

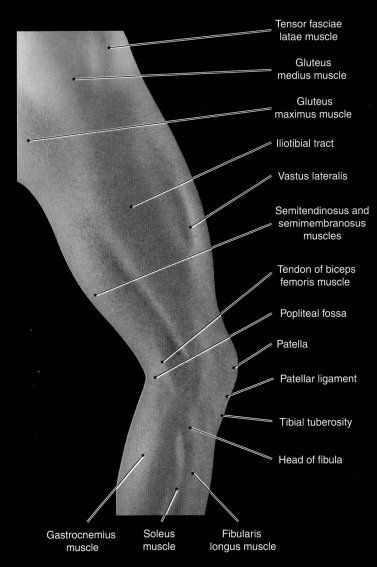

Tensor fasciae
latae muscle

Gluteus
medius muscle

Gluteus
maximus muscle

Iliotibial tract

Vastus lateralis

Semitendinosus and
semimembranosus
muscles

Tendon of biceps
femoris muscle

Popliteal fossa

Patella

Patellar ligament

Tibial tuberosity

Head of fibula

Gastrocnemius
muscle

Soleus
muscle

Fibularis
longus muscle

PLATE **69a** SURFACE ANATOMY OF THE THIGH,
LATERAL VIEW

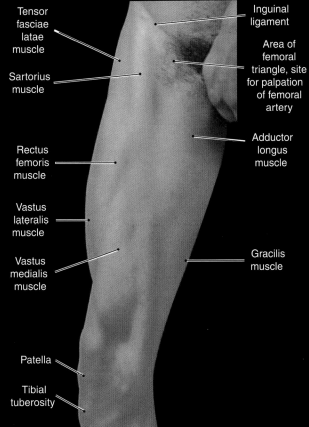

Tensor
fasciae
latae
muscle

Sartorius
muscle

Rectus
femoris
muscle

Vastus
lateralis
muscle

Vastus
medialis
muscle

Patella

Tibial
tuberosity

Inguinal
ligament

Area of
femoral
triangle, site
for palpation
of femoral
artery

Adductor
longus
muscle

Gracilis
muscle

PLATE **69b** SURFACE ANATOMY OF RIGHT THIGH,
ANTERIOR VIEW

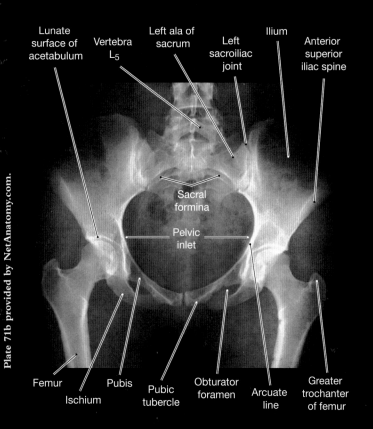

PLATE 70a **DISSECTION OF THE RIGHT INGUINAL REGION, MALE**

Inguinal ligament

Femoral artery

Deep inguinal lymph nodes

Great saphenous vein

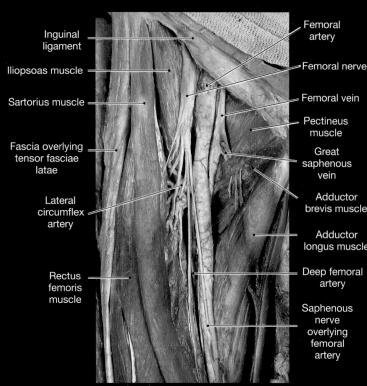

PLATE 70b **SUPERFICIAL DISSECTION OF RIGHT THIGH, ANTERIOR VIEW**

Inguinal ligament

Iliopsoas muscle

Sartorius muscle

Fascia overlying tensor fasciae latae

Lateral circumflex artery

Rectus femoris muscle

Femoral artery

Femoral nerve

Femoral vein

Pectineus muscle

Great saphenous vein

Adductor brevis muscle

Adductor longus muscle

Deep femoral artery

Saphenous nerve overlying femoral artery

Plate 71b provided by NetAnatomy.com.

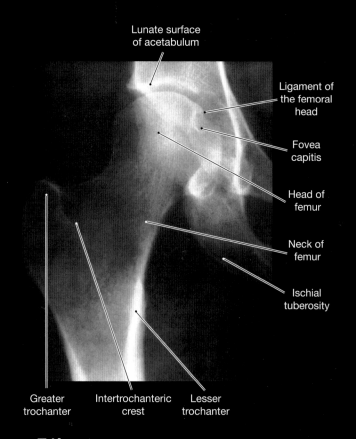

PLATE 71a **X-RAY OF PELVIS AND PROXIMAL FEMORA, ANTERIOR-POSTERIOR PROJECTION**

Lunate surface of acetabulum

Vertebra L_5

Left ala of sacrum

Left sacroiliac joint

Ilium

Anterior superior iliac spine

Sacral formina

Pelvic inlet

Femur

Ischium

Pubis

Pubic tubercle

Obturator foramen

Arcuate line

Greater trochanter of femur

PLATE 71b **X-RAY OF THE RIGHT HIP JOINT, ANTERIOR-POSTERIOR PROJECTION**

Lunate surface of acetabulum

Ligament of the femoral head

Fovea capitis

Head of femur

Neck of femur

Ischial tuberosity

Greater trochanter

Intertrochanteric crest

Lesser trochanter

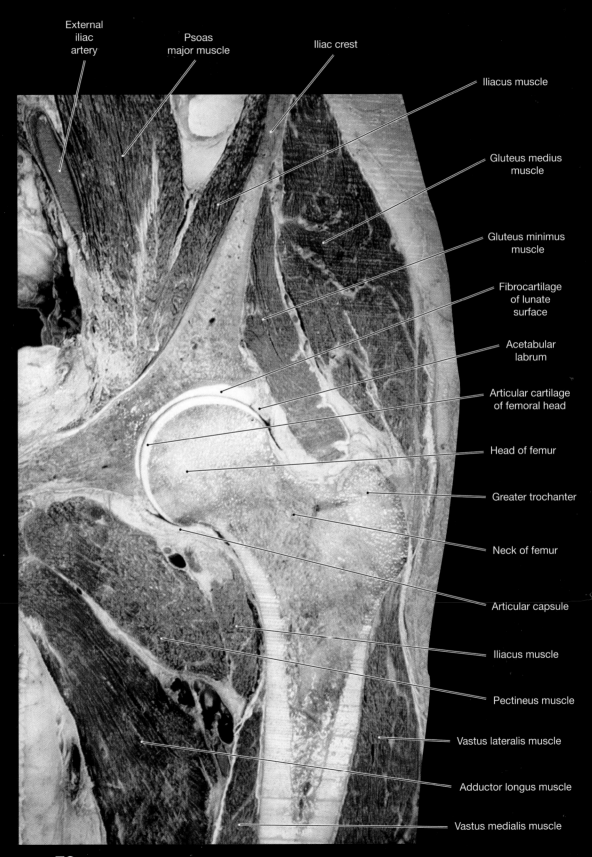

External
iliac
artery

Psoas
major muscle

Iliac crest

Iliacus muscle

Gluteus medius
muscle

Gluteus minimus
muscle

Fibrocartilage
of lunate
surface

Acetabular
labrum

Articular cartilage
of femoral head

Head of femur

Greater trochanter

Neck of femur

Articular capsule

Iliacus muscle

Pectineus muscle

Vastus lateralis muscle

Adductor longus muscle

Vastus medialis muscle

PLATE **72a** CORONAL SECTION THROUGH THE LEFT HIP

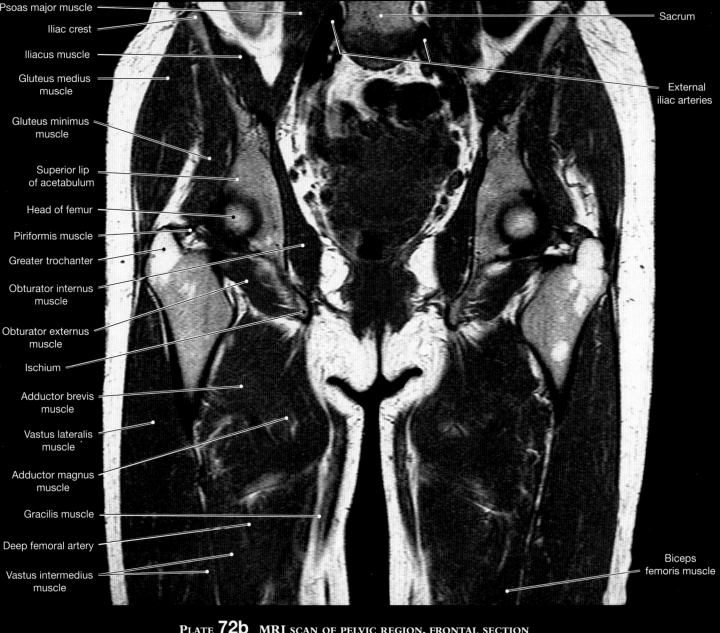

Psoas major muscle

Iliac crest

Iliacus muscle

Gluteus medius muscle

Gluteus minimus muscle

Superior lip of acetabulum

Head of femur

Piriformis muscle

Greater trochanter

Obturator internus muscle

Obturator externus muscle

Ischium

Adductor brevis muscle

Vastus lateralis muscle

Adductor magnus muscle

Gracilis muscle

Deep femoral artery

Vastus intermedius muscle

Sacrum

External iliac arteries

Biceps femoris muscle

PLATE **72b** MRI SCAN OF PELVIC REGION, FRONTAL SECTION

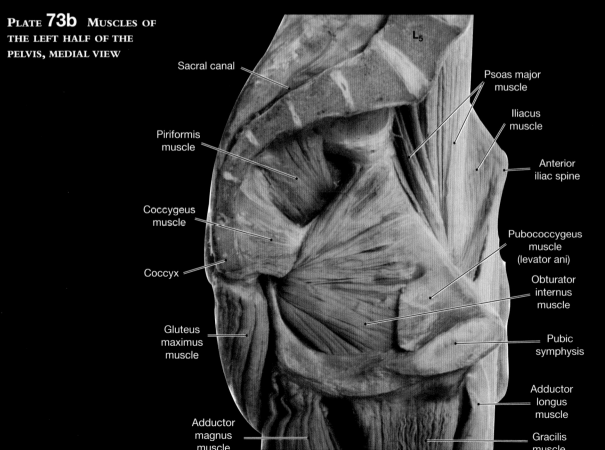

Iliac crest

Psoas major muscle

Iliacus muscle

Anterior superior iliac spine

Piriformis muscle

Inguinal ligament

Coccygeus muscle

Sartorius muscle

Pectineus muscle

Tensor fasciae latae muscle

Pubic tubercle

Adductor brevis muscle

Adductor longus muscle

Vastus lateralis muscle

Gracilis muscle

Rectus femoris muscle

L_5

Sacral canal

Psoas major muscle

Iliacus muscle

Piriformis muscle

Anterior iliac spine

Coccygeus muscle

Coccyx

Pubococcygeus muscle (levator ani)

Obturator internus muscle

Gluteus maximus muscle

Pubic symphysis

Adductor longus muscle

Adductor magnus muscle

Gracilis muscle

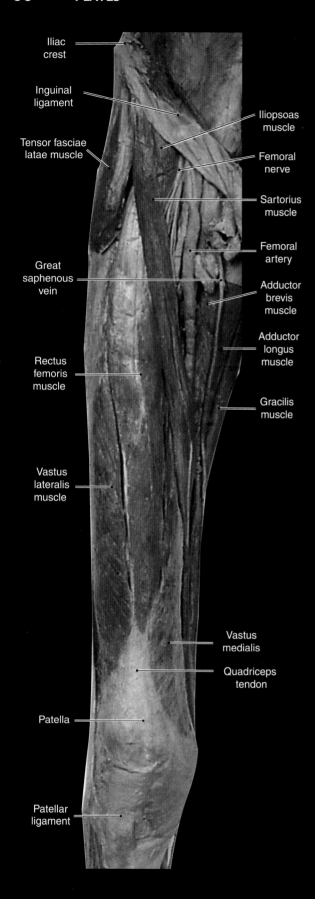

Iliac
crest

Inguinal
ligament

Tensor fasciae
latae muscle

Great
saphenous
vein

Rectus
femoris
muscle

Vastus
lateralis
muscle

Vastus
medialis

Patella

Patellar
ligament

Iliopsoas
muscle

Femoral
nerve

Sartorius
muscle

Femoral
artery

Adductor
brevis
muscle

Adductor
longus
muscle

Gracilis
muscle

Quadriceps
tendon

**PLATE 74 SUPERFICIAL DISSECTION OF RIGHT LOWER
LIMB, ANTERIOR VIEW**

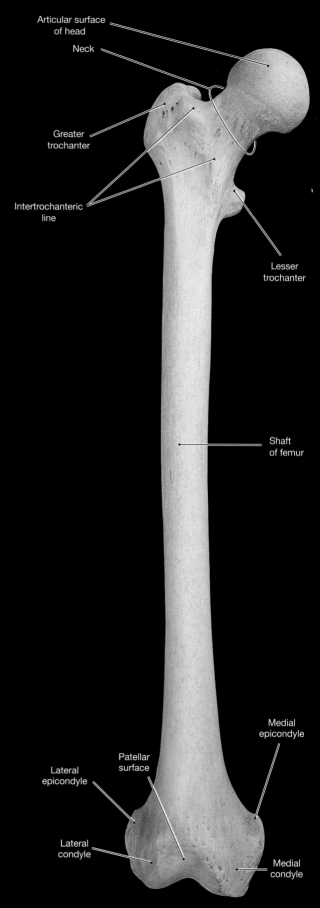

Articular surface
of head

Neck

Greater
trochanter

Intertrochanteric
line

Lesser
trochanter

Shaft
of femur

Medial
epicondyle

Patellar
surface

Lateral
epicondyle

Lateral
condyle

Medial
condyle

PLATE 75a RIGHT FEMUR, ANTERIOR VIEW

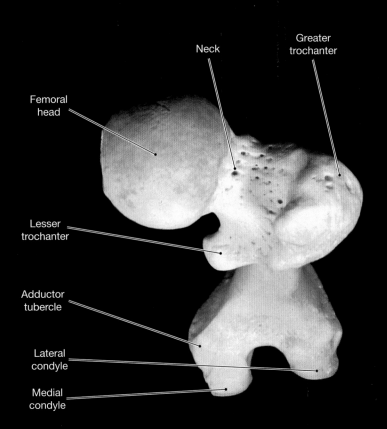

Femoral
head

Neck

Greater
trochanter

Lesser
trochanter

Adductor
tubercle

Lateral
condyle

Medial
condyle

PLATE 75b RIGHT FEMUR, SUPERIOR VIEW

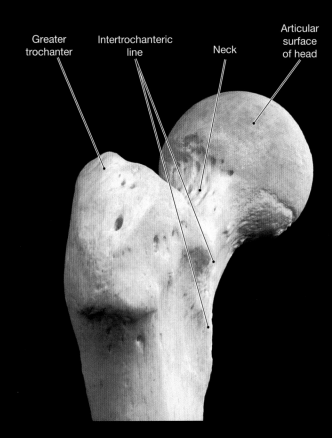

Greater
trochanter

Intertrochanteric
line

Neck

Articular
surface
of head

PLATE 75c HEAD OF RIGHT FEMUR, LATERAL VIEW

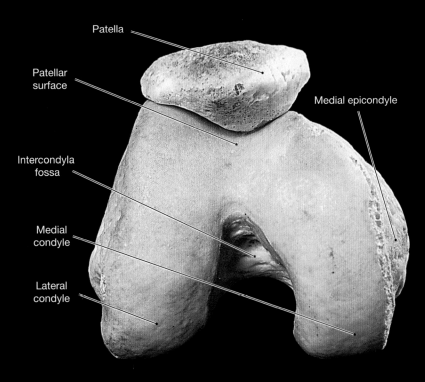

Patella

Patellar
surface

Medial epicondyle

Intercondyla
fossa

Medial
condyle

Lateral
condyle

PLATE 75d RIGHT FEMUR, INFERIOR VIEW

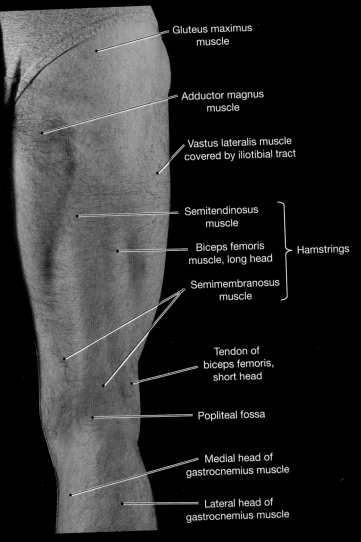

Gluteus maximus
muscle

Adductor magnus
muscle

Vastus lateralis muscle
covered by iliotibial tract

Semitendinosus
muscle

Biceps femoris
muscle, long head

} Hamstrings

Semimembranosus
muscle

Tendon of
biceps femoris,
short head

Popliteal fossa

Medial head of
gastrocnemius muscle

Lateral head of
gastrocnemius muscle

PLATE 76a SURFACE ANATOMY OF RIGHT THIGH,
POSTERIOR VIEW

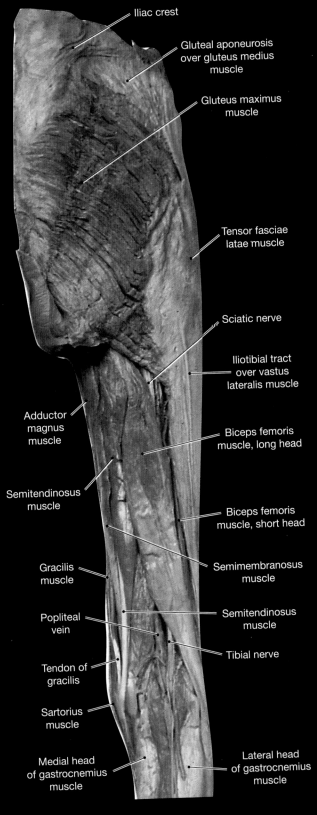

Iliac crest

Gluteal aponeurosis
over gluteus medius
muscle

Gluteus maximus
muscle

Tensor fasciae
latae muscle

Sciatic nerve

Iliotibial tract
over vastus
lateralis muscle

Biceps femoris
muscle, long head

Adductor
magnus
muscle

Biceps femoris
muscle, short head

Semitendinosus
muscle

Semimembranosus
muscle

Gracilis
muscle

Semitendinosus
muscle

Popliteal
vein

Tibial nerve

Tendon of
gracilis

Sartorius
muscle

Medial head
of gastrocnemius
muscle

Lateral head
of gastrocnemius
muscle

PLATE 76b SUPERFICIAL DISSECTION OF RIGHT
HIP AND THIGH, POSTERIOR VIEW

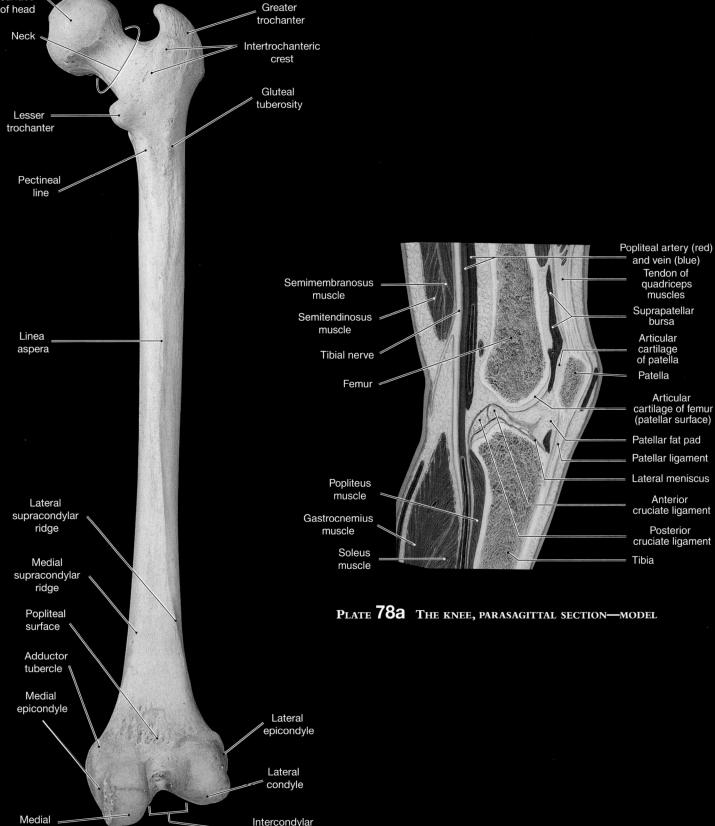

of head

Neck

Greater
trochanter

Intertrochanteric
crest

Lesser
trochanter

Gluteal
tuberosity

Pectineal
line

Linea
aspera

Lateral
supracondylar
ridge

Medial
supracondylar
ridge

Popliteal
surface

Adductor
tubercle

Medial
epicondyle

Lateral
epicondyle

Lateral
condyle

Medial

Intercondylar

Semimembranosus
muscle

Semitendinosus
muscle

Tibial nerve

Femur

Popliteus
muscle

Gastrocnemius
muscle

Soleus
muscle

Popliteal artery (red)
and vein (blue)

Tendon of
quadriceps
muscles

Suprapatellar
bursa

Articular
cartilage
of patella

Patella

Articular
cartilage of femur
(patellar surface)

Patellar fat pad

Patellar ligament

Lateral meniscus

Anterior
cruciate ligament

Posterior
cruciate ligament

Tibia

PLATE **78a** THE KNEE, PARASAGITTAL SECTION—MODEL

78b

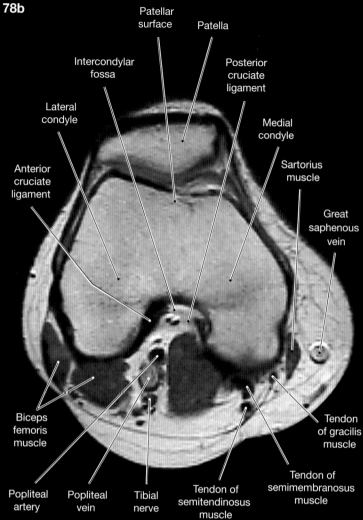

Patellar surface

Patella

Intercondylar fossa

Posterior cruciate ligament

Lateral condyle

Medial condyle

Anterior cruciate ligament

Sartorius muscle

Great saphenous vein

Biceps femoris muscle

Tendon of gracilis muscle

Popliteal artery

Popliteal vein

Tibial nerve

Tendon of semitendinosus muscle

Tendon of semimembranosus muscle

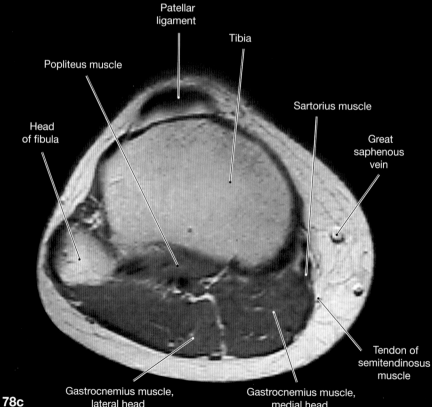

Patellar ligament

Tibia

Popliteus muscle

Head of fibula

Sartorius muscle

Great saphenous vein

Tendon of semitendinosus muscle

Gastrocnemius muscle, lateral head

Gastrocnemius muscle, medial head

PLATES **78b–c** MRI SCANS OF THE
RIGHT KNEE, HORIZONTAL SECTIONS,
SUPERIOR TO INFERIOR SEQUENCE

78c

78d

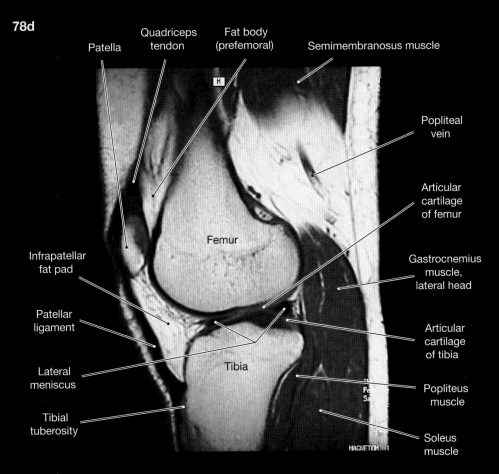

Patella
Quadriceps tendon
Fat body (prefemoral)
Semimembranosus muscle
Popliteal vein
Articular cartilage of femur
Gastrocnemius muscle, lateral head
Articular cartilage of tibia
Popliteus muscle
Soleus muscle
Infrapatellar fat pad
Patellar ligament
Lateral meniscus
Tibial tuberosity
Femur
Tibia

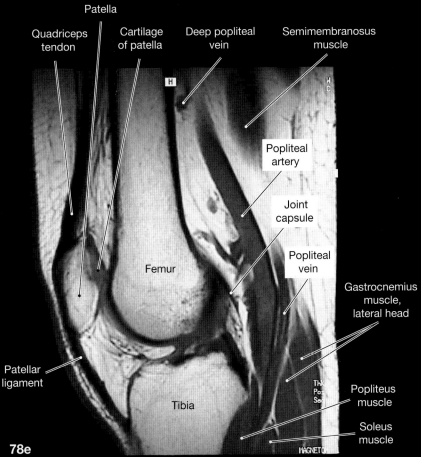

Patella
Quadriceps tendon
Cartilage of patella
Deep popliteal vein
Semimembranosus muscle
Popliteal artery
Joint capsule
Popliteal vein
Gastrocnemius muscle, lateral head
Popliteus muscle
Soleus muscle
Patellar ligament
Femur
Tibia

78e

PLATES **78d–e** MRI SCANS OF THE RIGHT KNEE, PARASAGITTAL SECTIONS, LATERAL TO MEDIAL SEQUENCE

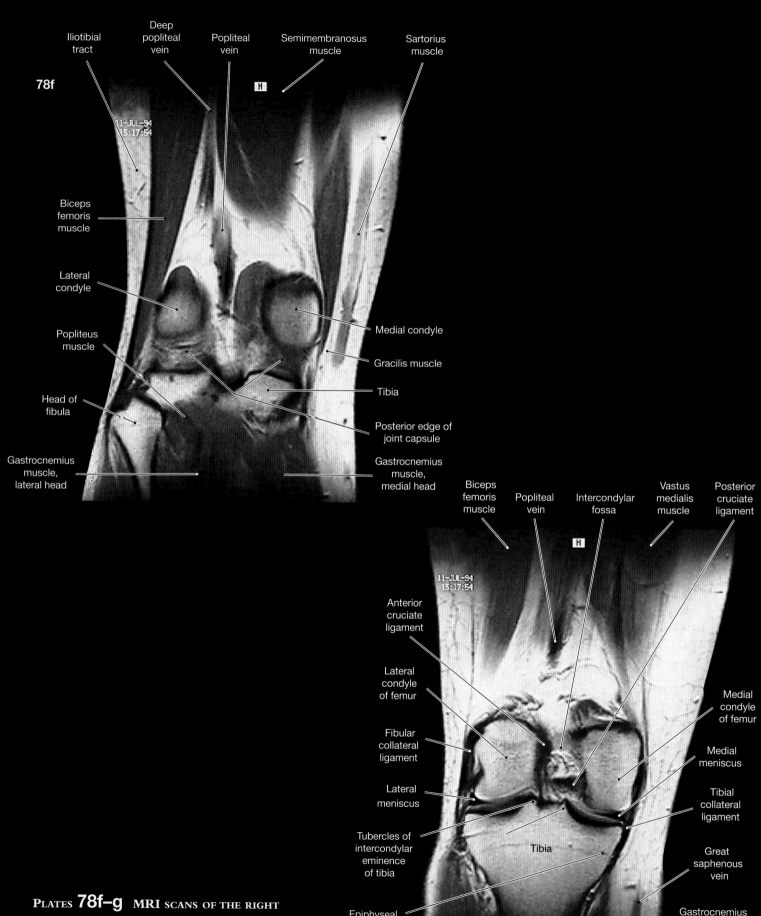

78f

Iliotibial tract

Deep popliteal vein

Popliteal vein

Semimembranosus muscle

Sartorius muscle

Biceps femoris muscle

Lateral condyle

Popliteus muscle

Head of fibula

Gastrocnemius muscle, lateral head

Medial condyle

Gracilis muscle

Tibia

Posterior edge of joint capsule

Gastrocnemius muscle, medial head

Biceps femoris muscle

Popliteal vein

Intercondylar fossa

Vastus medialis muscle

Posterior cruciate ligament

Anterior cruciate ligament

Lateral condyle of femur

Fibular collateral ligament

Lateral meniscus

Tubercles of intercondylar eminence of tibia

Epiphyseal line

Medial condyle of femur

Medial meniscus

Tibial collateral ligament

Great saphenous vein

Gastrocnemius muscle, medial head

Tibia

PLATES 78f–g MRI SCANS OF THE RIGHT KNEE, FRONTAL SECTIONS, POSTERIOR TO ANTERIOR SEQUENCE

78g

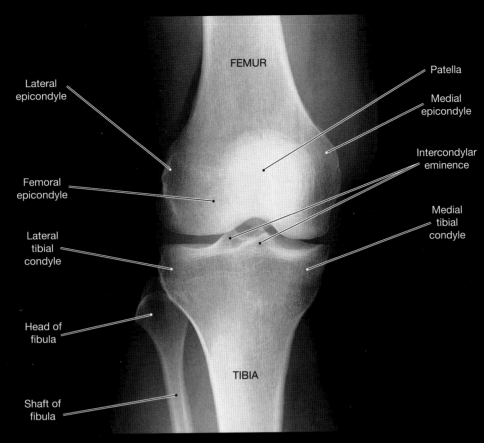

FEMUR

Lateral
epicondyle

Patella

Medial
epicondyle

Femoral
epicondyle

Intercondylar
eminence

Medial
tibial
condyle

Lateral
tibial
condyle

Head of
fibula

TIBIA

Shaft of
fibula

PLATE **78h**   X-RAY OF THE EXTENDED RIGHT KNEE, ANTERIOR-POSTERIOR PROJECTION

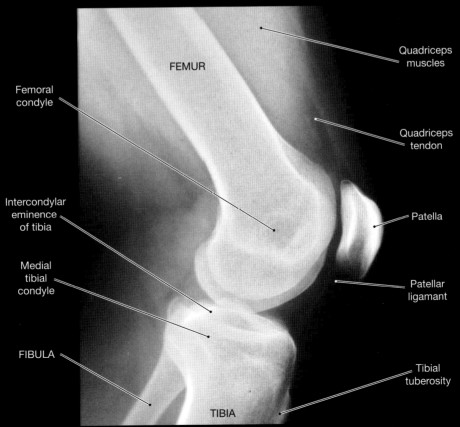

FEMUR

Quadriceps
muscles

Femoral
condyle

Quadriceps
tendon

Intercondylar
eminence
of tibia

Patella

Medial
tibial
condyle

Patellar
ligamant

FIBULA

Tibial
tuberosity

TIBIA

PLATE **78i**   X-RAY OF THE PARTIALLY FLEXED RIGHT KNEE, LATERAL PROJECTION

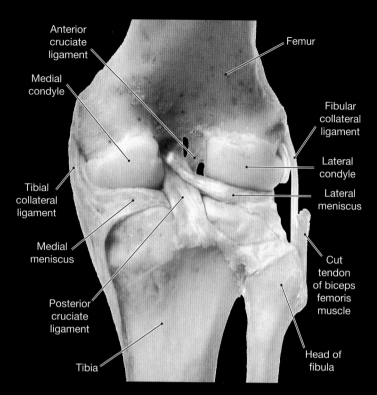

Anterior cruciate ligament

Femur

Medial condyle

Fibular collateral ligament

Lateral condyle

Tibial collateral ligament

Lateral meniscus

Medial meniscus

Posterior cruciate ligament

Cut tendon of biceps femoris muscle

Tibia

Head of fibula

PLATE **79a** EXTENDED RIGHT KNEE, POSTERIOR VIEW

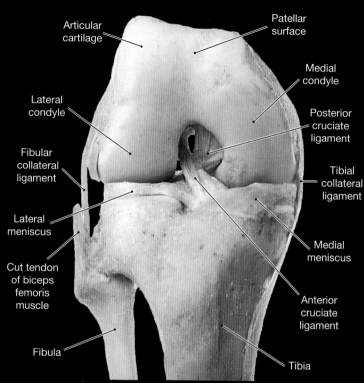

Articular cartilage

Patellar surface

Medial condyle

Lateral condyle

Posterior cruciate ligament

Fibular collateral ligament

Tibial collateral ligament

Lateral meniscus

Medial meniscus

Cut tendon of biceps femoris muscle

Anterior cruciate ligament

Fibula

Tibia

PLATE **79b** EXTENDED RIGHT KNEE, ANTERIOR VIEW

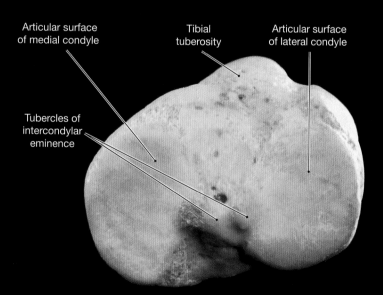

Articular surface of medial condyle

Tibial tuberosity

Articular surface of lateral condyle

Tubercles of intercondylar eminence

PLATE **80a** PROXIMAL END OF RIGHT TIBIA, SUPERIOR VIEW

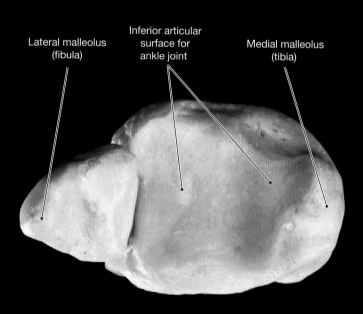

Lateral malleolus (fibula)

Inferior articular surface for ankle joint

Medial malleolus (tibia)

PLATE **80b** DISTAL END OF TIBIA AND FIBULA, INFERIOR VIEW

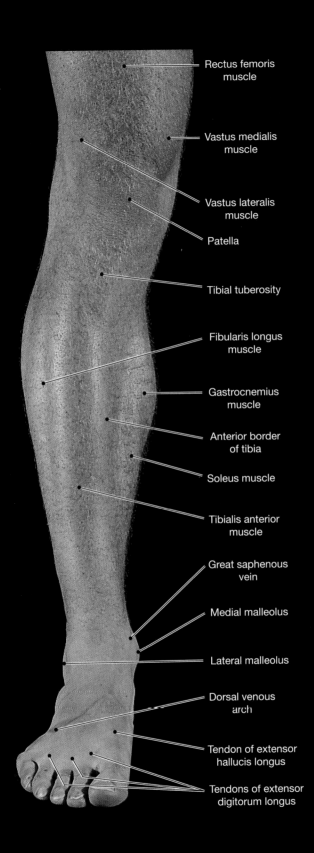

Rectus femoris muscle

Vastus medialis muscle

Vastus lateralis muscle

Patella

Tibial tuberosity

Fibularis longus muscle

Gastrocnemius muscle

Anterior border of tibia

Soleus muscle

Tibialis anterior muscle

Great saphenous vein

Medial malleolus

Lateral malleolus

Dorsal venous arch

Tendon of extensor hallucis longus

Tendons of extensor digitorum longus

PLATE 81a SURFACE ANATOMY OF THE RIGHT LEG AND FOOT, ANTERIOR VIEW

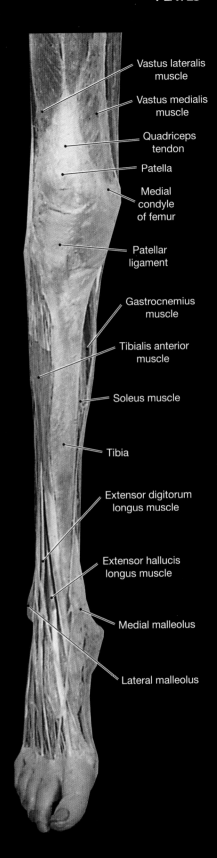

Vastus lateralis muscle

Vastus medialis muscle

Quadriceps tendon

Patella

Medial condyle of femur

Patellar ligament

Gastrocnemius muscle

Tibialis anterior muscle

Soleus muscle

Tibia

Extensor digitorum longus muscle

Extensor hallucis longus muscle

Medial malleolus

Lateral malleolus

PLATE 81b SUPERFICIAL DISSECTION OF THE RIGHT LEG AND FOOT, ANTERIOR VIEW

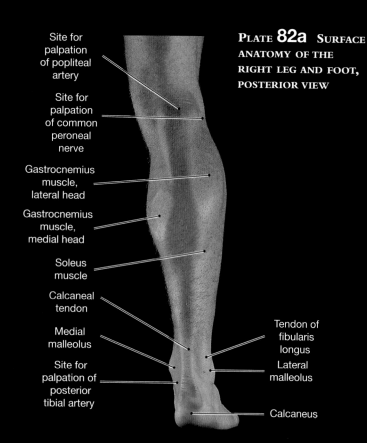

Site for palpation of popliteal artery

Site for palpation of common peroneal nerve

Gastrocnemius muscle, lateral head

Gastrocnemius muscle, medial head

Soleus muscle

Calcaneal tendon

Medial malleolus

Site for palpation of posterior tibial artery

Tendon of fibularis longus

Lateral malleolus

Calcaneus

PLATE **82a** SURFACE ANATOMY OF THE RIGHT LEG AND FOOT, POSTERIOR VIEW

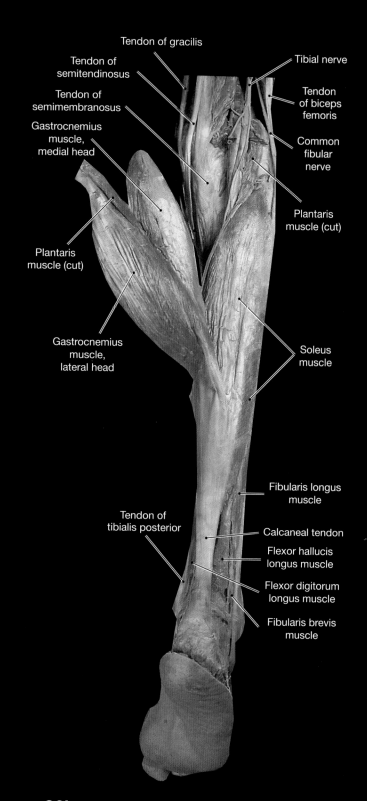

Tendon of gracilis

Tendon of semitendinosus

Tendon of semimembranosus

Gastrocnemius muscle, medial head

Plantaris muscle (cut)

Gastrocnemius muscle, lateral head

Tibial nerve

Tendon of biceps femoris

Common fibular nerve

Plantaris muscle (cut)

Soleus muscle

Fibularis longus muscle

Tendon of tibialis posterior

Calcaneal tendon

Flexor hallucis longus muscle

Flexor digitorum longus muscle

Fibularis brevis muscle

PLATE **82b** SUPERFICIAL DISSECTION OF THE RIGHT LEG AND FOOT, POSTERIOR VIEW

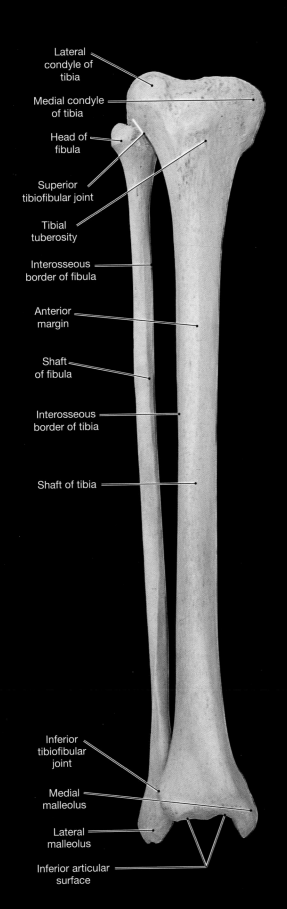

Lateral
condyle of
tibia

Medial condyle
of tibia

Head of
fibula

Superior
tibiofibular joint

Tibial
tuberosity

Interosseous
border of fibula

Anterior
margin

Shaft
of fibula

Interosseous
border of tibia

Shaft of tibia

Inferior
tibiofibular
joint

Medial
malleolus

Lateral
malleolus

Inferior articular
surface

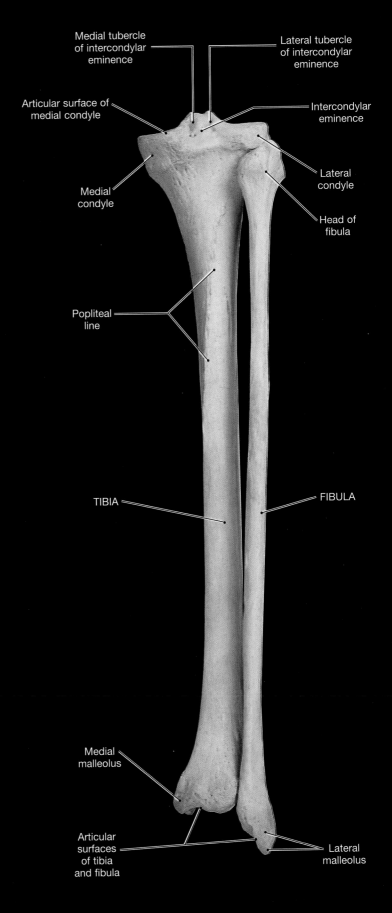

Medial tubercle
of intercondylar
eminence

Lateral tubercle
of intercondylar
eminence

Articular surface of
medial condyle

Intercondylar
eminence

Medial
condyle

Lateral
condyle

Head of
fibula

Popliteal
line

TIBIA

FIBULA

Medial
malleolus

Articular
surfaces
of tibia
and fibula

Lateral
malleolus

PLATE **83a** Tibia and fibula, anterior view

PLATE **83b** Tibia and fibula, posterior view

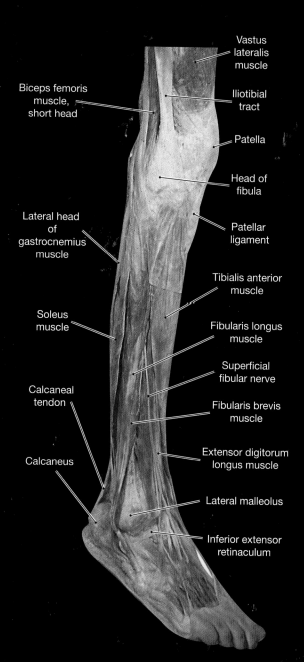

Vastus lateralis muscle

Biceps femoris muscle, short head

Iliotibial tract

Patella

Head of fibula

Lateral head of gastrocnemius muscle

Patellar ligament

Tibialis anterior muscle

Soleus muscle

Fibularis longus muscle

Superficial fibular nerve

Calcaneal tendon

Fibularis brevis muscle

Extensor digitorum longus muscle

Calcaneus

Lateral malleolus

Inferior extensor retinaculum

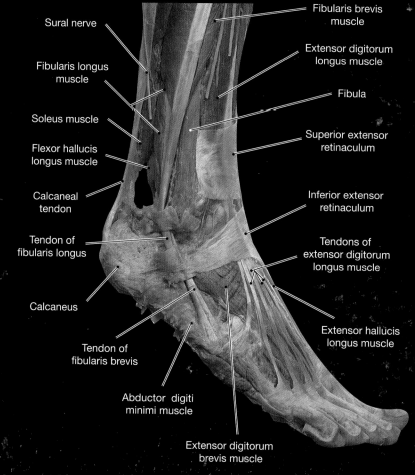

Sural nerve

Fibularis longus muscle

Soleus muscle

Flexor hallucis longus muscle

Calcaneal tendon

Tendon of fibularis longus

Calcaneus

Tendon of fibularis brevis

Abductor digiti minimi muscle

Extensor digitorum brevis muscle

Fibularis brevis muscle

Extensor digitorum longus muscle

Fibula

Superior extensor retinaculum

Inferior extensor retinaculum

Tendons of extensor digitorum longus muscle

Extensor hallucis longus muscle

PLATE **84a** SUPERFICIAL DISSECTION OF THE RIGHT LEG AND FOOT, ANTEROLATERAL VIEW

PLATE **84b** SUPERFICIAL DISSECTION OF THE RIGHT FOOT, LATERAL VIEW

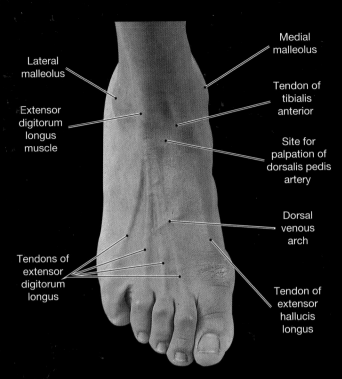

Lateral
malleolus

Extensor
digitorum
longus
muscle

Medial
malleolus

Tendon of
tibialis
anterior

Site for
palpation of
dorsalis pedis
artery

Dorsal
venous
arch

Tendons of
extensor
digitorum
longus

Tendon of
extensor
hallucis
longus

PLATE 85a SURFACE ANATOMY OF THE RIGHT FOOT,
SUPERIOR VIEW

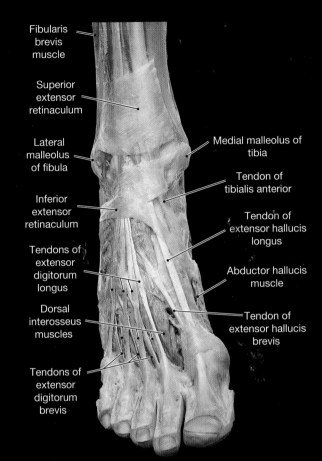

Fibularis
brevis
muscle

Superior
extensor
retinaculum

Lateral
malleolus
of fibula

Inferior
extensor
retinaculum

Tendons of
extensor
digitorum
longus

Dorsal
interosseus
muscles

Tendons of
extensor
digitorum
brevis

Medial malleolus of
tibia

Tendon of
tibialis anterior

Tendon of
extensor hallucis
longus

Abductor hallucis
muscle

Tendon of
extensor hallucis
brevis

PLATE 85b SUPERFICIAL DISSECTION OF THE RIGHT
FOOT, SUPERIOR VIEW

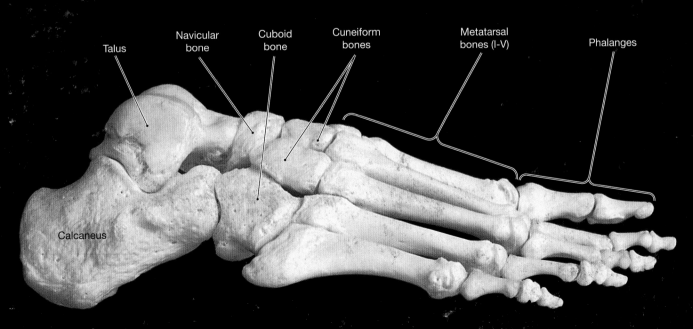

Talus

Navicular
bone

Cuboid
bone

Cuneiform
bones

Metatarsal
bones (I-V)

Phalanges

Calcaneus

PLATE 86a BONES OF THE RIGHT FOOT, LATERAL VIEW

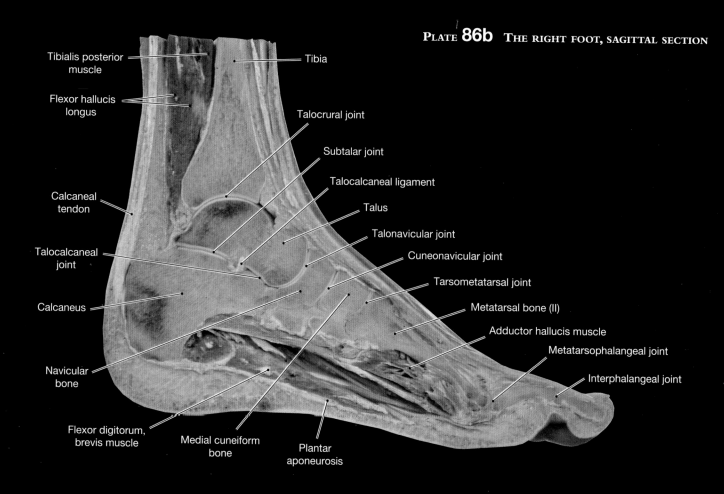

Tibialis posterior muscle

Flexor hallucis longus

Calcaneal tendon

Talocalcaneal joint

Calcaneus

Navicular bone

Flexor digitorum, brevis muscle

Medial cuneiform bone

Plantar aponeurosis

Tibia

Talocrural joint

Subtalar joint

Talocalcaneal ligament

Talus

Talonavicular joint

Cuneonavicular joint

Tarsometatarsal joint

Metatarsal bone (II)

Adductor hallucis muscle

Metatarsophalangeal joint

Interphalangeal joint

PLATE **86c** MRI SCAN OF THE RIGHT ANKLE, SAGITTAL SECTION

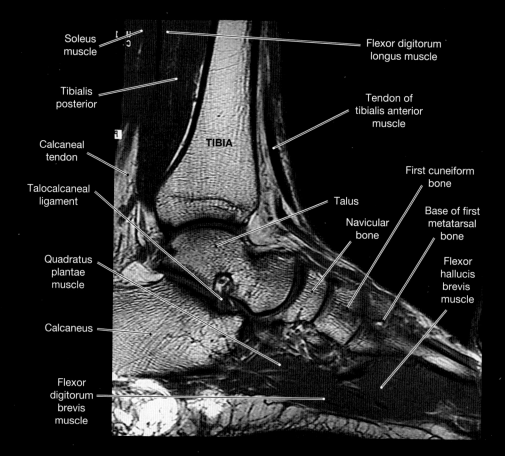

Soleus muscle

Tibialis posterior

Calcaneal tendon

Talocalcaneal ligament

Quadratus plantae muscle

Calcaneus

Flexor digitorum brevis muscle

TIBIA

Flexor digitorum longus muscle

Tendon of tibialis anterior muscle

First cuneiform bone

Talus

Navicular bone

Base of first metatarsal bone

Flexor hallucis brevis muscle

PLATE **87a** ANKLE AND FOOT,
POSTERIOR VIEW

Tendon of
flexor digitorum
longus muscle

Flexor hallucis
longus muscle

Medial
malleolus

Site for
palpation of
posterior
tibial artery

Calcaneus

Tendon of
fibularis longus
muscle

Calcaneal
tendon

Lateral
malleolus

Tendon of
fibularis brevis
muscle

Base of fifth
metatarsal
bone

Tibia

Talus

Medial
malleolus

Deltoid
ligament

Talocalcaneal
ligament

Calcaneocuboid
joint

Fibula

Talocrural
(ankle) joint

Lateral
malleolus

Calcaneus

Cuboid
bone

PLATE **87b** FRONTAL SECTION THROUGH THE RIGHT FOOT,
POSTERIOR VIEW

Extensor
digitorum
longus muscle

Tibia

Lateral
malleolus
of fibula

Talus

Calcaneus

Tendon of
fibularis
longus
muscle

Abductor
digiti minimi
muscle

Medial malleolus
of tibia

Tendon of
tibialis posterior
muscle

Deltoid
ligament

Tendon of
flexor digitorum
longus muscle

Tendon of
flexor hallucis
longus muscle

Plantar artery

Abductor
hallucis muscle

Quadratus
plantae muscle

Flexor
digitorum
brevis muscle

PLATE **87c** MRI SCAN OF THE RIGHT ANKLE, FRONTAL
SECTION

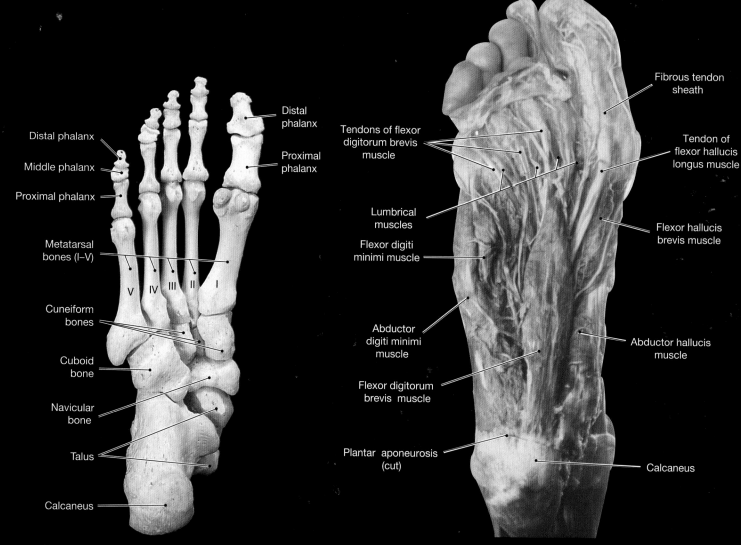

Distal phalanx

Distal phalanx

Middle phalanx

Proximal phalanx

Metatarsal bones (I–V)

Cuneiform bones

Cuboid bone

Navicular bone

Talus

Calcaneus

Distal phalanx

Proximal phalanx

V   IV   III   II   I

Fibrous tendon sheath

Tendons of flexor digitorum brevis muscle

Lumbrical muscles

Flexor digiti minimi muscle

Abductor digiti minimi muscle

Flexor digitorum brevis muscle

Plantar aponeurosis (cut)

Tendon of flexor hallucis longus muscle

Flexor hallucis brevis muscle

Abductor hallucis muscle

Calcaneus

PLATE 88 BONES OF THE RIGHT FOOT, INFERIOR (PLANTAR) VIEW

PLATE 89 SUPERFICIAL DISSECTION OF THE RIGHT FOOT, PLANTAR VIEW

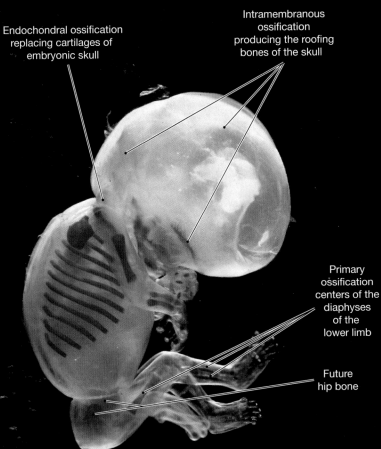

Endochondral ossification replacing cartilages of embryonic skull

Intramembranous ossification producing the roofing bones of the skull

PLATE **90a** SKELETON OF FETUS AFTER **10** WEEKS OF DEVELOPMENT

Primary ossification centers of the diaphyses of the lower limb

Future hip bone

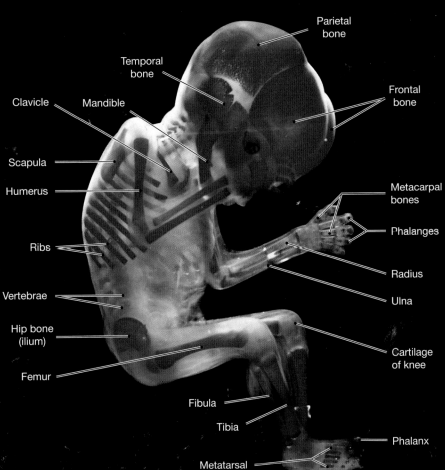

Parietal bone

Temporal bone

Clavicle

Mandible

Frontal bone

Scapula

Humerus

Metacarpal bones

Phalanges

Ribs

Radius

Vertebrae

Ulna

Hip bone (ilium)

Cartilage of knee

Femur

Fibula

Tibia

Phalanx

Metatarsal bones

PLATE **90b** SKELETON OF FETUS AFTER **16** WEEKS OF DEVELOPMENT

Embryology Summaries

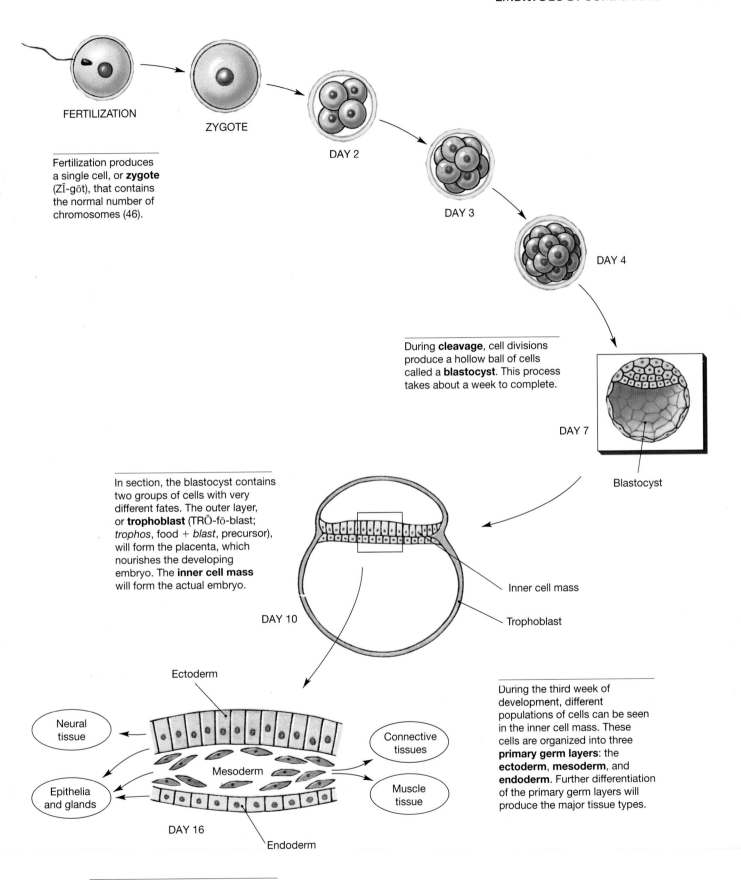

FERTILIZATION

ZYGOTE

Fertilization produces a single cell, or **zygote** (ZĪ-gōt), that contains the normal number of chromosomes (46).

DAY 2

DAY 3

DAY 4

During **cleavage**, cell divisions produce a hollow ball of cells called a **blastocyst**. This process takes about a week to complete.

DAY 7

Blastocyst

In section, the blastocyst contains two groups of cells with very different fates. The outer layer, or **trophoblast** (TRŌ-fō-blast; *trophos*, food + *blast*, precursor), will form the placenta, which nourishes the developing embryo. The **inner cell mass** will form the actual embryo.

DAY 10

Inner cell mass

Trophoblast

Ectoderm

Neural tissue

Mesoderm

Epithelia and glands

DAY 16

Endoderm

Connective tissues

Muscle tissue

During the third week of development, different populations of cells can be seen in the inner cell mass. These cells are organized into three **primary germ layers**: the **ectoderm**, **mesoderm**, and **endoderm**. Further differentiation of the primary germ layers will produce the major tissue types.

All three germ layers participate in the formation of functional organs and organ systems. Their interactions will be detailed in later Embryology Summaries dealing with specific systems.

EMBRYOLOGY SUMMARY 1: THE FORMATION OF TISSUES

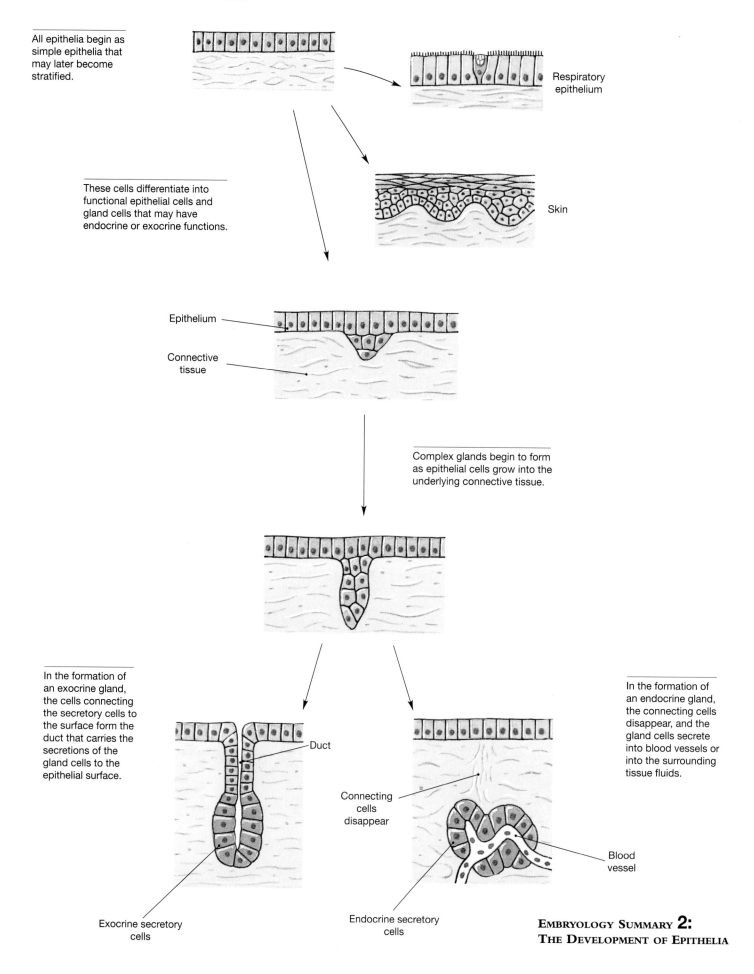

All epithelia begin as simple epithelia that may later become stratified.

Respiratory epithelium

These cells differentiate into functional epithelial cells and gland cells that may have endocrine or exocrine functions.

Skin

Epithelium

Connective tissue

Complex glands begin to form as epithelial cells grow into the underlying connective tissue.

In the formation of an exocrine gland, the cells connecting the secretory cells to the surface form the duct that carries the secretions of the gland cells to the epithelial surface.

Duct

In the formation of an endocrine gland, the connecting cells disappear, and the gland cells secrete into blood vessels or into the surrounding tissue fluids.

Connecting cells disappear

Blood vessel

Exocrine secretory cells

Endocrine secretory cells

EMBRYOLOGY SUMMARY 2:
THE DEVELOPMENT OF EPITHELIA

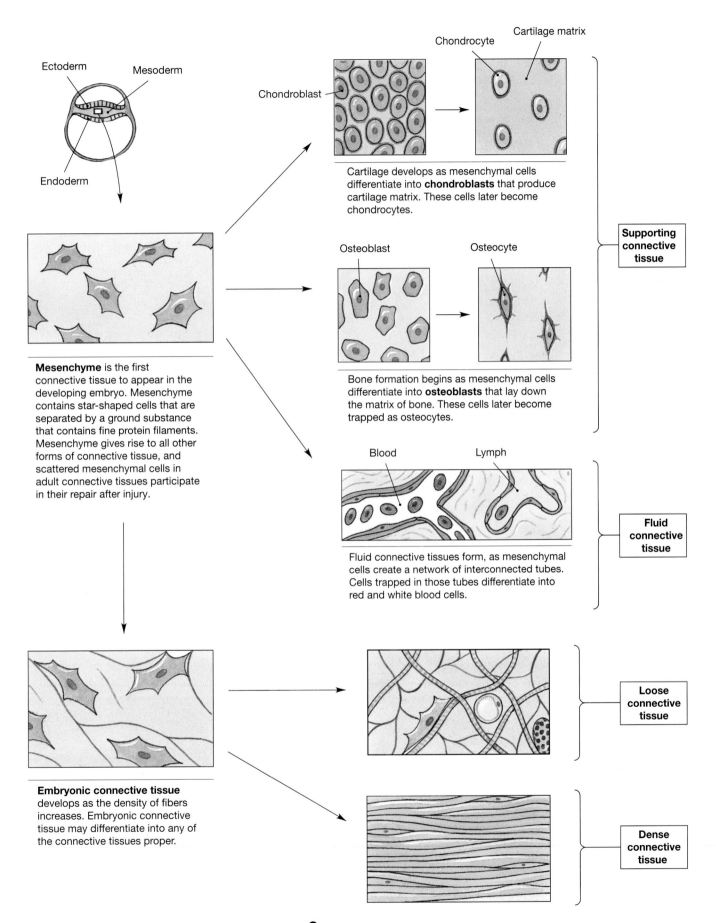

Ectoderm   Mesoderm

Endoderm

Chondroblast

Chondrocyte

Cartilage matrix

Cartilage develops as mesenchymal cells differentiate into **chondroblasts** that produce cartilage matrix. These cells later become chondrocytes.

Osteoblast

Osteocyte

Bone formation begins as mesenchymal cells differentiate into **osteoblasts** that lay down the matrix of bone. These cells later become trapped as osteocytes.

Blood

Lymph

Fluid connective tissues form, as mesenchymal cells create a network of interconnected tubes. Cells trapped in those tubes differentiate into red and white blood cells.

Mesenchyme is the first connective tissue to appear in the developing embryo. Mesenchyme contains star-shaped cells that are separated by a ground substance that contains fine protein filaments. Mesenchyme gives rise to all other forms of connective tissue, and scattered mesenchymal cells in adult connective tissues participate in their repair after injury.

Embryonic connective tissue develops as the density of fibers increases. Embryonic connective tissue may differentiate into any of the connective tissues proper.

Supporting connective tissue

Fluid connective tissue

Loose connective tissue

Dense connective tissue

EMBRYOLOGY SUMMARY **3:** THE ORIGINS OF CONNECTIVE TISSUES

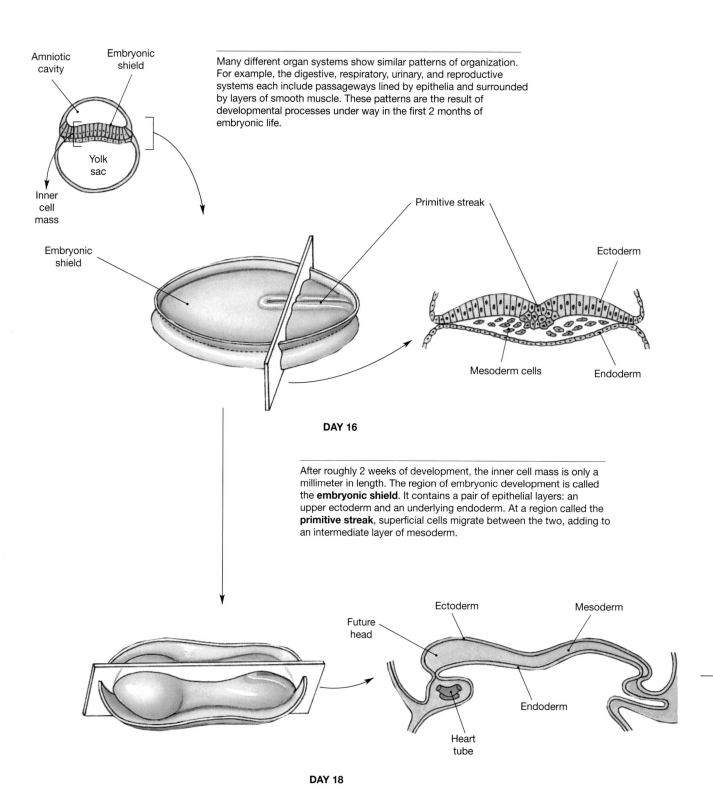

Amniotic cavity

Embryonic shield

Yolk sac

Inner cell mass

Many different organ systems show similar patterns of organization. For example, the digestive, respiratory, urinary, and reproductive systems each include passageways lined by epithelia and surrounded by layers of smooth muscle. These patterns are the result of developmental processes under way in the first 2 months of embryonic life.

Embryonic shield

Primitive streak

Ectoderm

Mesoderm cells

Endoderm

DAY 16

After roughly 2 weeks of development, the inner cell mass is only a millimeter in length. The region of embryonic development is called the **embryonic shield**. It contains a pair of epithelial layers: an upper ectoderm and an underlying endoderm. At a region called the **primitive streak**, superficial cells migrate between the two, adding to an intermediate layer of mesoderm.

Ectoderm

Mesoderm

Future head

Endoderm

Heart tube

DAY 18

By day 18, the embryo has begun to lift off the surface of the embryonic shield. The heart and many blood vessels have already formed, well ahead of the other organ systems. Unless otherwise noted, discussions of organ system development in later chapters will begin at this stage.

EMBRYOLOGY SUMMARY 4: THE DEVELOPMENT OF ORGAN SYSTEMS

DERIVATIVES OF PRIMARY GERM LAYERS	
Ectoderm Forms:	Epidermis and epidermal derivatives of the integumentary system, including hair follicles, nails, and glands communicating with the skin surface (sweat, milk, and sebum) Lining of the mouth, salivary glands, nasal passageways, and anus Nervous system, including brain and spinal cord Portions of endocrine system (pituitary gland and medulla of adrenal gland) Portions of skull, pharyngeal arches, and teeth
Mesoderm Forms:	Dermis of the skin Lining of the body cavities (pleural, pericardial, peritoneal) Muscular, skeletal, cardiovascular, and lymphatic systems Kidneys and part of the urinary tract Gonads and most of the reproductive tracts Connective tissues supporting all organ systems Portions of endocrine system (parts of adrenal glands and endocrine tissues of the reproductive tracts)
Endoderm Forms:	Most of the digestive system: epithelium (except mouth and anus), exocrine glands (except salivary glands), the liver and pancreas Most of the respiratory system: epithelium (except nasal passageways) and mucous glands Portions of urinary and reproductive systems (ducts and the stem cells that produce gametes) Portions of endocrine system (thymus, thyroid gland, parathyroid glands, and pancreas)

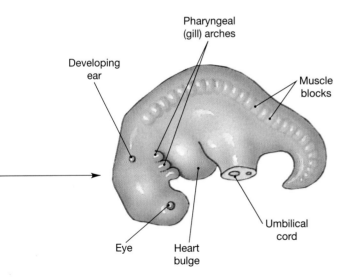

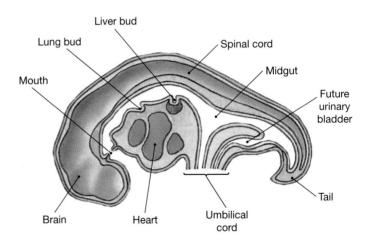

DAY 28

After 1 month, you can find the beginnings of all major organ systems. The role of each of the primary germ layers in the formation of organs is summarized in the accompanying table; details are given in later Embryology Summaries.

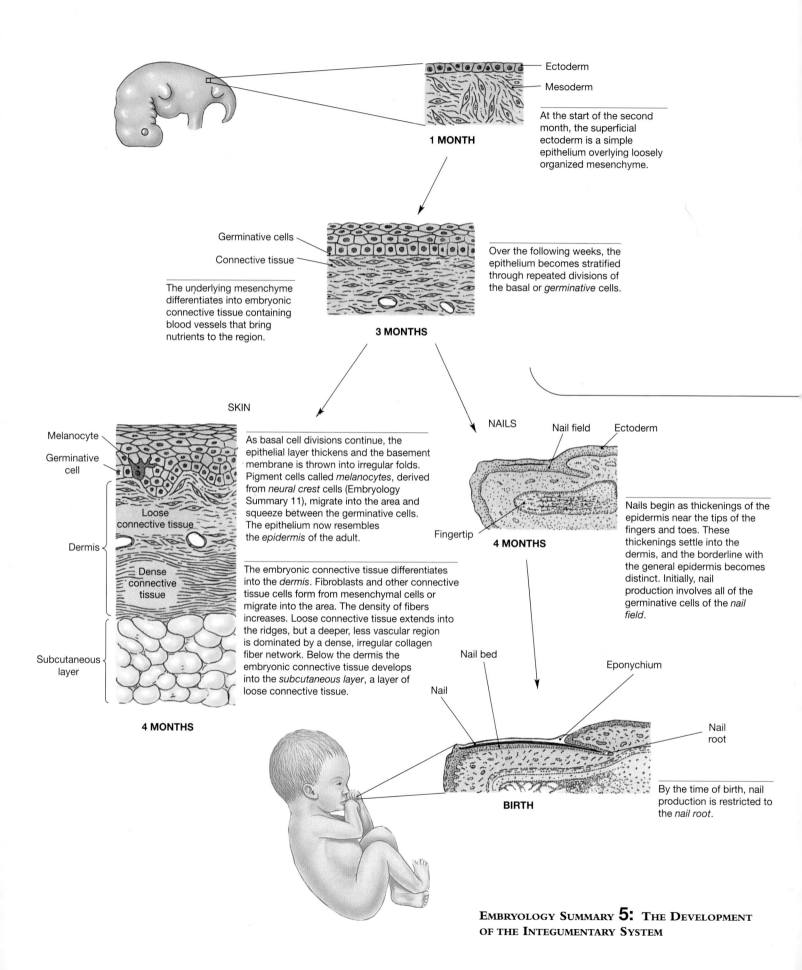

1 MONTH

Ectoderm

Mesoderm

At the start of the second month, the superficial ectoderm is a simple epithelium overlying loosely organized mesenchyme.

Germinative cells

Connective tissue

The underlying mesenchyme differentiates into embryonic connective tissue containing blood vessels that bring nutrients to the region.

3 MONTHS

Over the following weeks, the epithelium becomes stratified through repeated divisions of the basal or *germinative* cells.

SKIN

Melanocyte

Germinative cell

Loose connective tissue

Dermis

Dense connective tissue

Subcutaneous layer

4 MONTHS

As basal cell divisions continue, the epithelial layer thickens and the basement membrane is thrown into irregular folds. Pigment cells called *melanocytes*, derived from *neural crest* cells (Embryology Summary 11), migrate into the area and squeeze between the germinative cells. The epithelium now resembles the *epidermis* of the adult.

The embryonic connective tissue differentiates into the *dermis*. Fibroblasts and other connective tissue cells form from mesenchymal cells or migrate into the area. The density of fibers increases. Loose connective tissue extends into the ridges, but a deeper, less vascular region is dominated by a dense, irregular collagen fiber network. Below the dermis the embryonic connective tissue develops into the *subcutaneous layer*, a layer of loose connective tissue.

NAILS

Nail field

Ectoderm

Fingertip

4 MONTHS

Nails begin as thickenings of the epidermis near the tips of the fingers and toes. These thickenings settle into the dermis, and the borderline with the general epidermis becomes distinct. Initially, nail production involves all of the germinative cells of the *nail field*.

Nail bed

Nail

Eponychium

Nail root

BIRTH

By the time of birth, nail production is restricted to the *nail root*.

EMBRYOLOGY SUMMARY 5: THE DEVELOPMENT OF THE INTEGUMENTARY SYSTEM

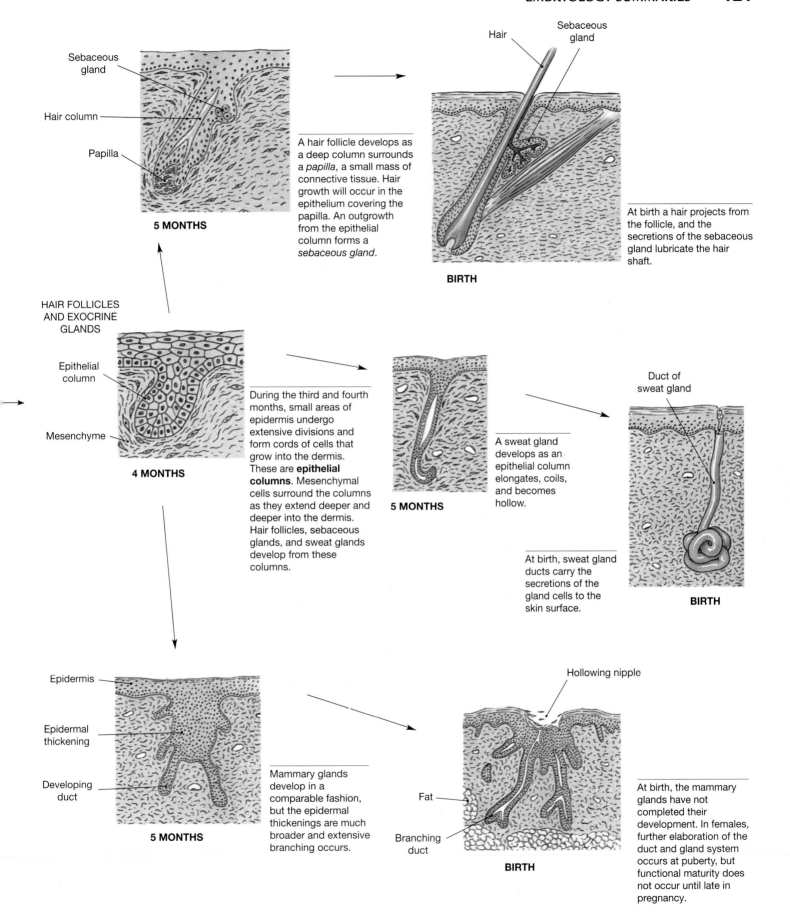

Sebaceous gland

Hair column

Papilla

5 MONTHS

A hair follicle develops as a deep column surrounds a *papilla*, a small mass of connective tissue. Hair growth will occur in the epithelium covering the papilla. An outgrowth from the epithelial column forms a *sebaceous gland*.

Hair

Sebaceous gland

BIRTH

At birth a hair projects from the follicle, and the secretions of the sebaceous gland lubricate the hair shaft.

HAIR FOLLICLES AND EXOCRINE GLANDS

Epithelial column

Mesenchyme

4 MONTHS

During the third and fourth months, small areas of epidermis undergo extensive divisions and form cords of cells that grow into the dermis. These are **epithelial columns**. Mesenchymal cells surround the columns as they extend deeper and deeper into the dermis. Hair follicles, sebaceous glands, and sweat glands develop from these columns.

5 MONTHS

A sweat gland develops as an epithelial column elongates, coils, and becomes hollow.

Duct of sweat gland

At birth, sweat gland ducts carry the secretions of the gland cells to the skin surface.

BIRTH

Epidermis

Epidermal thickening

Developing duct

5 MONTHS

Mammary glands develop in a comparable fashion, but the epidermal thickenings are much broader and extensive branching occurs.

Hollowing nipple

Fat

Branching duct

BIRTH

At birth, the mammary glands have not completed their development. In females, further elaboration of the duct and gland system occurs at puberty, but functional maturity does not occur until late in pregnancy.

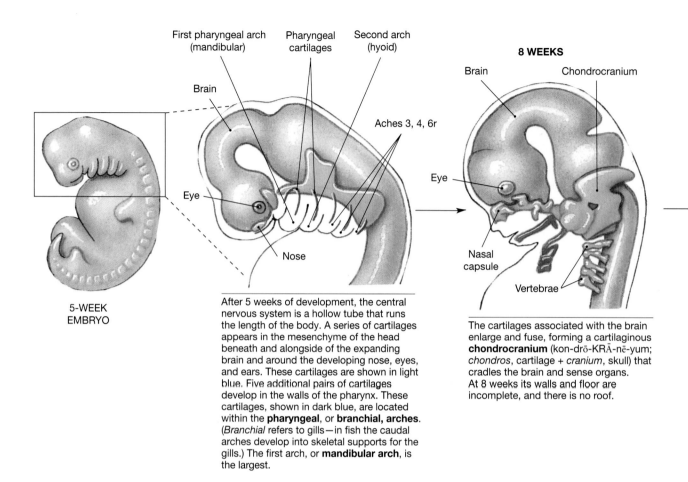

First pharyngeal arch (mandibular)

Pharyngeal cartilages

Second arch (hyoid)

Brain

Aches 3, 4, 6r

Eye

Nose

8 WEEKS

Brain

Chondrocranium

Eye

Nasal capsule

Vertebrae

5-WEEK EMBRYO

After 5 weeks of development, the central nervous system is a hollow tube that runs the length of the body. A series of cartilages appears in the mesenchyme of the head beneath and alongside of the expanding brain and around the developing nose, eyes, and ears. These cartilages are shown in light blue. Five additional pairs of cartilages develop in the walls of the pharynx. These cartilages, shown in dark blue, are located within the **pharyngeal**, or **branchial, arches**. (*Branchial* refers to gills—in fish the caudal arches develop into skeletal supports for the gills.) The first arch, or **mandibular arch**, is the largest.

The cartilages associated with the brain enlarge and fuse, forming a cartilaginous **chondrocranium** (kon-drō-KRĀ-nē-yum; *chondros*, cartilage + *cranium*, skull) that cradles the brain and sense organs. At 8 weeks its walls and floor are incomplete, and there is no roof.

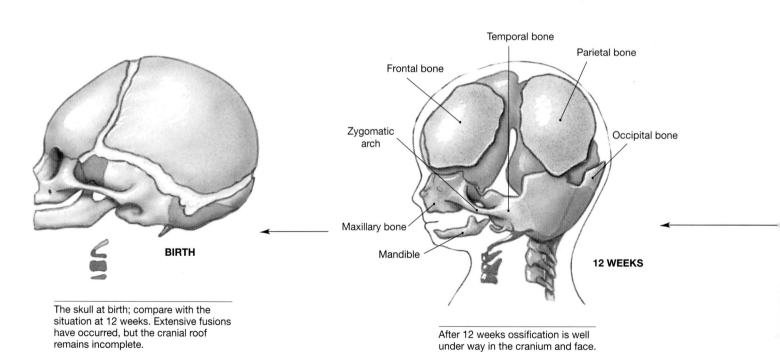

Temporal bone

Frontal bone

Parietal bone

Zygomatic arch

Occipital bone

Maxillary bone

Mandible

BIRTH

12 WEEKS

The skull at birth; compare with the situation at 12 weeks. Extensive fusions have occurred, but the cranial roof remains incomplete.

After 12 weeks ossification is well under way in the cranium and face.

EMBRYOLOGY SUMMARY 6: THE DEVELOPMENT OF THE SKULL

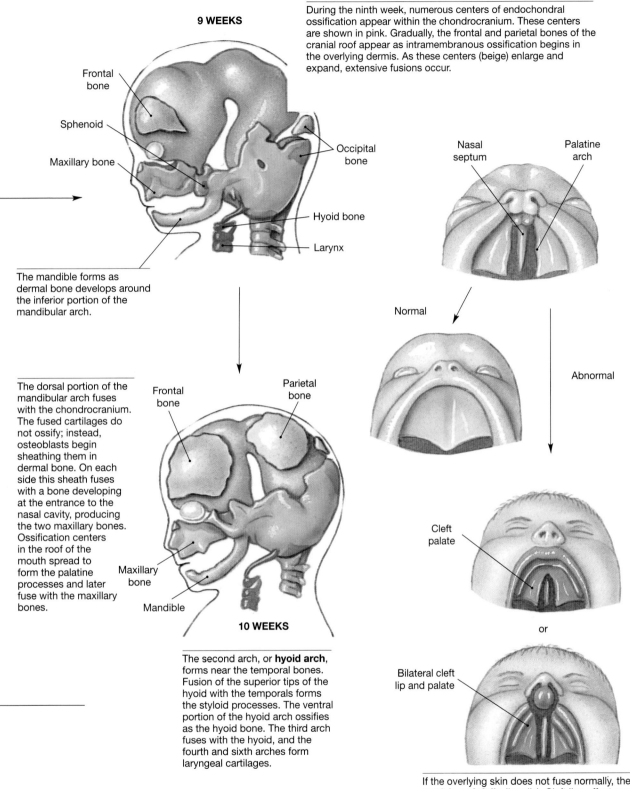

9 WEEKS

During the ninth week, numerous centers of endochondral ossification appear within the chondrocranium. These centers are shown in pink. Gradually, the frontal and parietal bones of the cranial roof appear as intramembranous ossification begins in the overlying dermis. As these centers (beige) enlarge and expand, extensive fusions occur.

Frontal bone

Sphenoid

Maxillary bone

Occipital bone

Hyoid bone

Larynx

The mandible forms as dermal bone develops around the inferior portion of the mandibular arch.

Nasal septum

Palatine arch

Normal

Abnormal

The dorsal portion of the mandibular arch fuses with the chondrocranium. The fused cartilages do not ossify; instead, osteoblasts begin sheathing them in dermal bone. On each side this sheath fuses with a bone developing at the entrance to the nasal cavity, producing the two maxillary bones. Ossification centers in the roof of the mouth spread to form the palatine processes and later fuse with the maxillary bones.

Frontal bone

Parietal bone

Maxillary bone

Mandible

10 WEEKS

The second arch, or **hyoid arch**, forms near the temporal bones. Fusion of the superior tips of the hyoid with the temporals forms the styloid processes. The ventral portion of the hyoid arch ossifies as the hyoid bone. The third arch fuses with the hyoid, and the fourth and sixth arches form laryngeal cartilages.

Cleft palate

or

Bilateral cleft lip and palate

If the overlying skin does not fuse normally, the result is a **cleft lip** (*harelip*). Cleft lips affect roughly one birth in a thousand. A split extending into the orbit and palate is called a **cleft palate**. Cleft palates are half as common as cleft lips. Both conditions can be corrected surgically.

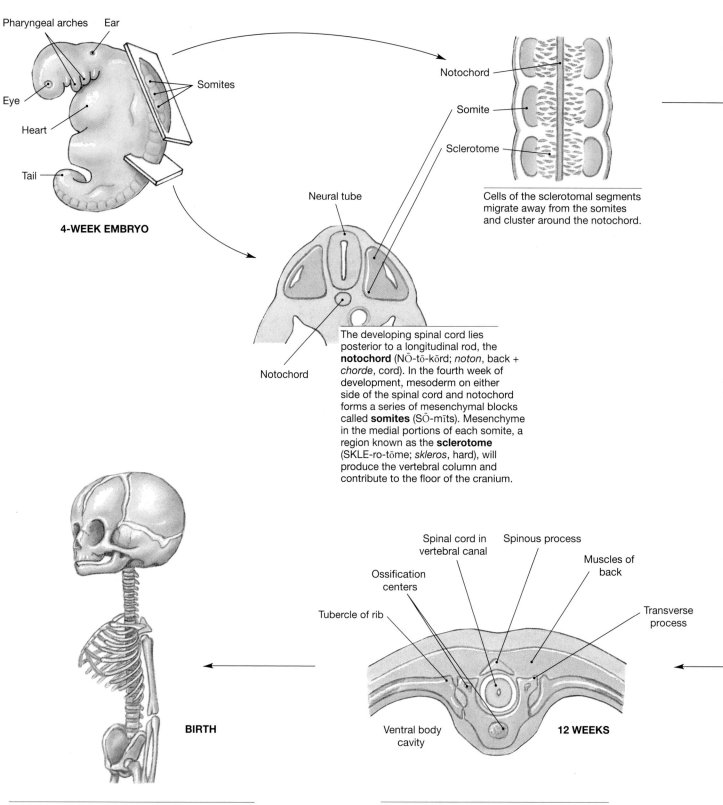

4-WEEK EMBRYO

- Pharyngeal arches
- Ear
- Eye
- Heart
- Tail
- Somites

Cells of the sclerotomal segments migrate away from the somites and cluster around the notochord.

- Notochord
- Somite
- Sclerotome
- Neural tube
- Notochord

The developing spinal cord lies posterior to a longitudinal rod, the **notochord** (NŌ-tō-kōrd; *noton*, back + *chorde*, cord). In the fourth week of development, mesoderm on either side of the spinal cord and notochord forms a series of mesenchymal blocks called **somites** (SŌ-mīts). Mesenchyme in the medial portions of each somite, a region known as the **sclerotome** (SKLE-ro-tōme; *skleros*, hard), will produce the vertebral column and contribute to the floor of the cranium.

12 WEEKS

- Spinal cord in vertebral canal
- Spinous process
- Muscles of back
- Ossification centers
- Tubercle of rib
- Transverse process
- Ventral body cavity

About the time the ribs separate from the vertebrae, ossification begins. Only the shortest ribs undergo complete ossification. In the rest, the distal portions remain cartilaginous, forming the costal cartilages. Several ossification centers appear in the sternum, but fusion gradually reduces the number.

BIRTH

At birth, the vertebrae and ribs are ossified, but many cartilaginous areas remain. For example, the anterior portions of the ribs remain cartilaginous. Additional growth will occur for many years; in vertebrae, the bases of the neural arches enlarge until ages 3–6, and the spinal processes and vertebral bodies grow until ages 18–25.

EMBRYOLOGY SUMMARY **7:** THE DEVELOPMENT OF THE VERTEBRAL COLUMN

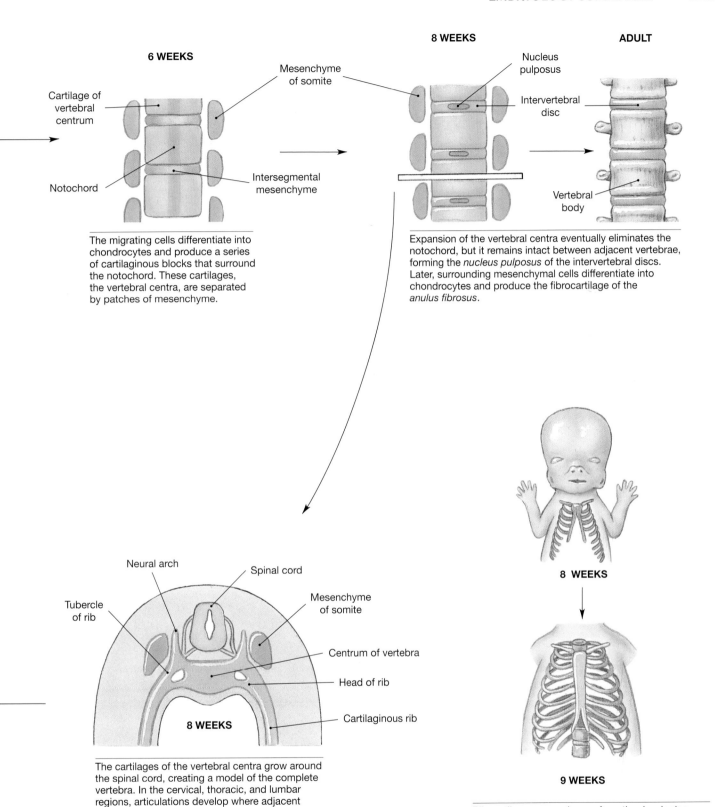

6 WEEKS

Cartilage of vertebral centrum

Mesenchyme of somite

Notochord

Intersegmental mesenchyme

The migrating cells differentiate into chondrocytes and produce a series of cartilaginous blocks that surround the notochord. These cartilages, the vertebral centra, are separated by patches of mesenchyme.

8 WEEKS

Nucleus pulposus

Intervertebral disc

ADULT

Vertebral body

Expansion of the vertebral centra eventually eliminates the notochord, but it remains intact between adjacent vertebrae, forming the *nucleus pulposus* of the intervertebral discs. Later, surrounding mesenchymal cells differentiate into chondrocytes and produce the fibrocartilage of the *anulus fibrosus*.

Neural arch

Tubercle of rib

Spinal cord

Mesenchyme of somite

Centrum of vertebra

Head of rib

Cartilaginous rib

8 WEEKS

The cartilages of the vertebral centra grow around the spinal cord, creating a model of the complete vertebra. In the cervical, thoracic, and lumbar regions, articulations develop where adjacent cartilaginous blocks come into contact. In the sacrum and coccyx, the cartilages fuse together.

8 WEEKS

9 WEEKS

Rib cartilages expand away from the developing transverse processes of the vertebrae. At first they are continuous, but by week 8 the ribs have separated from the vertebrae. Ribs form at every vertebra, but in the cervical, lumbar, sacral, and coccygeal regions, they remain small and later fuse with the growing vertebrae. The ribs of the thoracic vertebrae continue to enlarge, following the curvature of the body wall. When they reach the ventral midline, they fuse with the cartilages of the sternum.

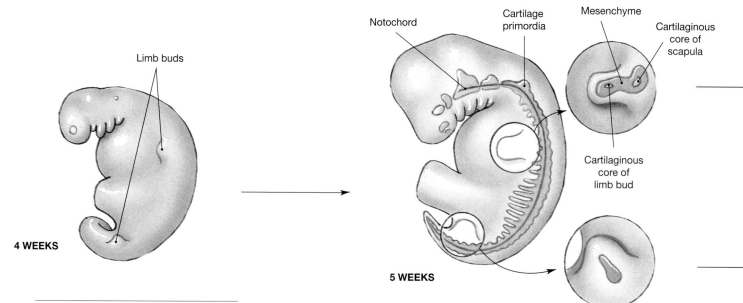

4 WEEKS

Notochord

Cartilage primordia

Mesenchyme

Cartilaginous core of scapula

Cartilaginous core of limb bud

Limb buds

5 WEEKS

In the fourth week of development, ridges appear along the flanks of the embryo, extending from just behind the throat to just before the anus. These ridges form as mesodermal cells congregate beneath the ectoderm of the flank. Mesoderm gradually accumulates at the end of each ridge, forming two pairs of limb buds.

After 5 weeks of development, the pectoral limb buds have a cartilaginous core and scapular cartilages are developing in the mesenchyme of the trunk.

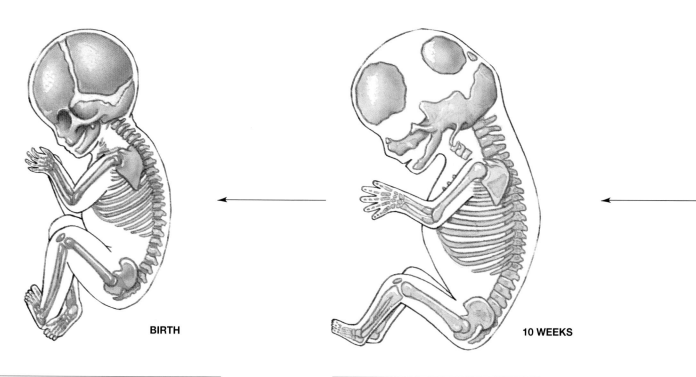

BIRTH

10 WEEKS

The skeleton of a newborn infant. Note the extensive areas of cartilage (blue) in the humeral head, in the wrist, between the bones of the palm and fingers, and in the coxae. Notice the appearance of the axial skeleton, with reference to Embryology Summaries 6 and 7.

Ossification in the embryonic skeleton after approximately 10 weeks of development. The shafts of the limb bones are undergoing rapid ossification, but the distal bones of the carpus and tarsus remain cartilaginous.

EMBRYOLOGY SUMMARY 8: THE DEVELOPMENT OF THE APPENDICULAR SKELETON

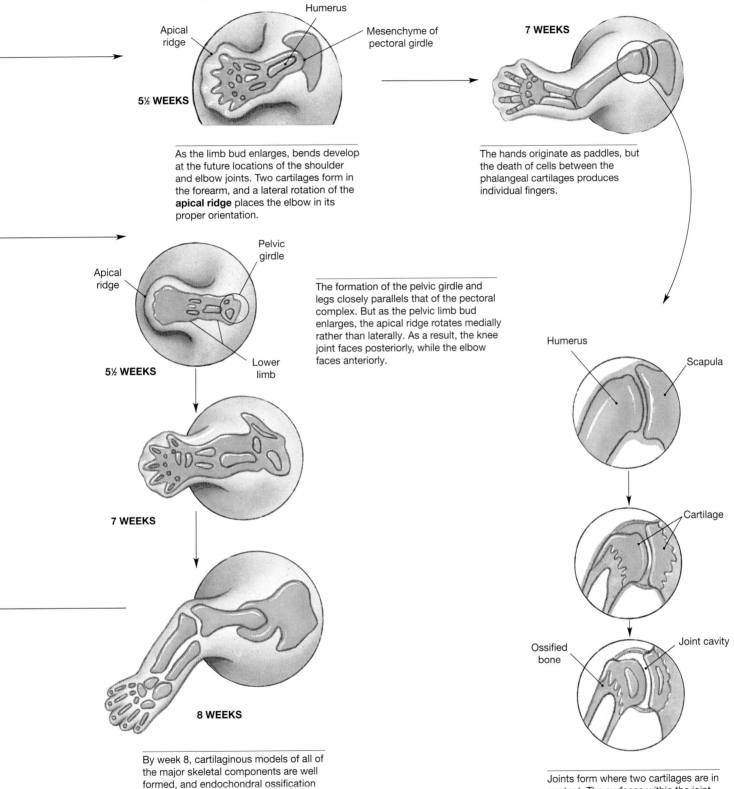

5½ WEEKS

Apical ridge

Humerus

Mesenchyme of pectoral girdle

As the limb bud enlarges, bends develop at the future locations of the shoulder and elbow joints. Two cartilages form in the forearm, and a lateral rotation of the **apical ridge** places the elbow in its proper orientation.

7 WEEKS

The hands originate as paddles, but the death of cells between the phalangeal cartilages produces individual fingers.

5½ WEEKS

Apical ridge

Pelvic girdle

Lower limb

The formation of the pelvic girdle and legs closely parallels that of the pectoral complex. But as the pelvic limb bud enlarges, the apical ridge rotates medially rather than laterally. As a result, the knee joint faces posteriorly, while the elbow faces anteriorly.

7 WEEKS

8 WEEKS

By week 8, cartilaginous models of all of the major skeletal components are well formed, and endochondral ossification begins in the future limb bones. Ossification of the ossa coxae begins at three separate centers that gradually enlarge.

Humerus

Scapula

Cartilage

Ossified bone

Joint cavity

Joints form where two cartilages are in contact. The surfaces within the joint cavity remain cartilaginous, while the rest of the bones undergo ossification.

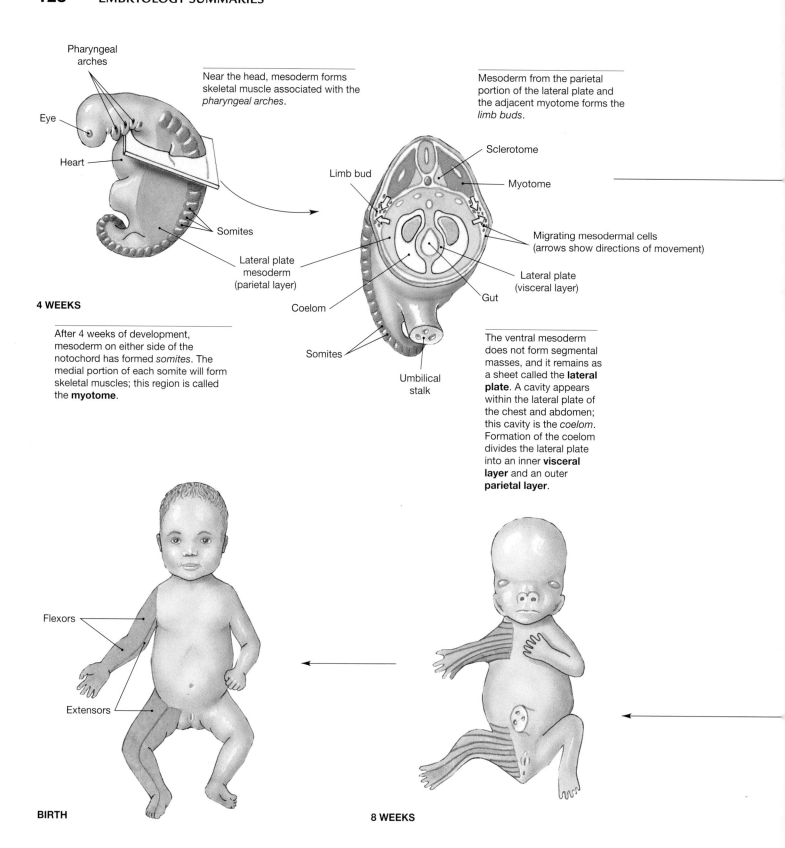

Pharyngeal arches

Eye

Heart

Near the head, mesoderm forms skeletal muscle associated with the *pharyngeal arches*.

Mesoderm from the parietal portion of the lateral plate and the adjacent myotome forms the *limb buds*.

Limb bud

Sclerotome

Myotome

Migrating mesodermal cells (arrows show directions of movement)

Lateral plate (visceral layer)

Somites

Lateral plate mesoderm (parietal layer)

Coelom

Gut

4 WEEKS

Somites

Umbilical stalk

After 4 weeks of development, mesoderm on either side of the notochord has formed *somites*. The medial portion of each somite will form skeletal muscles; this region is called the **myotome**.

The ventral mesoderm does not form segmental masses, and it remains as a sheet called the **lateral plate**. A cavity appears within the lateral plate of the chest and abdomen; this cavity is the *coelom*. Formation of the coelom divides the lateral plate into an inner **visceral layer** and an outer **parietal layer**.

Flexors

Extensors

BIRTH

8 WEEKS

Rotation of the arm and leg buds produces a change in the position of these masses relative to the body axis.

While the limb buds enlarge, additional myoblasts invade the limb from myotomal segments nearby. Lines indicate the boundaries between myotomes providing myoblasts to the limb.

EMBRYOLOGY SUMMARY 9: THE DEVELOPMENT OF THE MUSCULAR SYSTEM

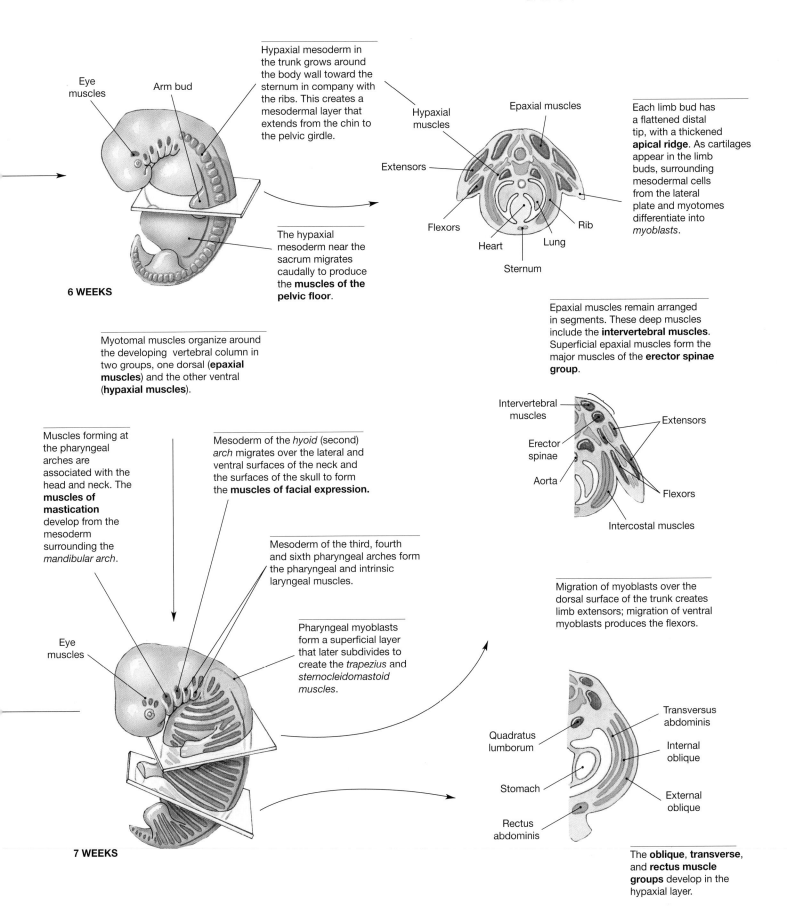

Eye muscles

Arm bud

Hypaxial mesoderm in the trunk grows around the body wall toward the sternum in company with the ribs. This creates a mesodermal layer that extends from the chin to the pelvic girdle.

The hypaxial mesoderm near the sacrum migrates caudally to produce the **muscles of the pelvic floor**.

6 WEEKS

Hypaxial muscles

Epaxial muscles

Extensors

Flexors

Heart

Lung

Rib

Sternum

Each limb bud has a flattened distal tip, with a thickened **apical ridge**. As cartilages appear in the limb buds, surrounding mesodermal cells from the lateral plate and myotomes differentiate into *myoblasts*.

Myotomal muscles organize around the developing vertebral column in two groups, one dorsal (**epaxial muscles**) and the other ventral (**hypaxial muscles**).

Epaxial muscles remain arranged in segments. These deep muscles include the **intervertebral muscles**. Superficial epaxial muscles form the major muscles of the **erector spinae group**.

Intervertebral muscles

Erector spinae

Aorta

Extensors

Flexors

Intercostal muscles

Muscles forming at the pharyngeal arches are associated with the head and neck. The **muscles of mastication** develop from the mesoderm surrounding the *mandibular arch*.

Mesoderm of the *hyoid* (second) *arch* migrates over the lateral and ventral surfaces of the neck and the surfaces of the skull to form the **muscles of facial expression.**

Mesoderm of the third, fourth and sixth pharyngeal arches form the pharyngeal and intrinsic laryngeal muscles.

Pharyngeal myoblasts form a superficial layer that later subdivides to create the *trapezius* and *sternocleidomastoid muscles*.

Migration of myoblasts over the dorsal surface of the trunk creates limb extensors; migration of ventral myoblasts produces the flexors.

Eye muscles

7 WEEKS

Quadratus lumborum

Stomach

Rectus abdominis

Transversus abdominis

Internal oblique

External oblique

The **oblique, transverse,** and **rectus muscle groups** develop in the hypaxial layer.

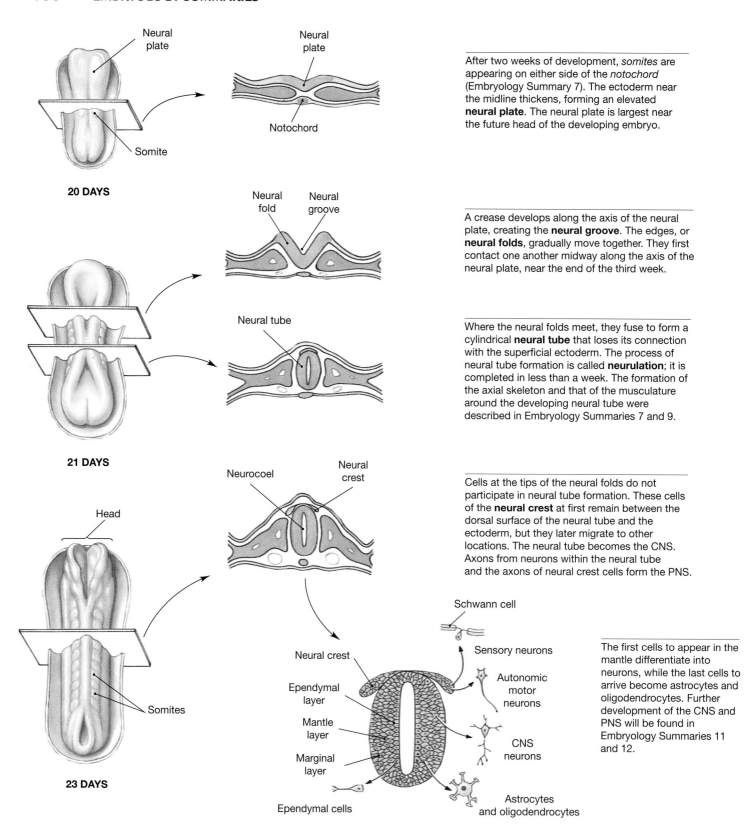

20 DAYS

After two weeks of development, *somites* are appearing on either side of the *notochord* (Embryology Summary 7). The ectoderm near the midline thickens, forming an elevated **neural plate**. The neural plate is largest near the future head of the developing embryo.

21 DAYS

A crease develops along the axis of the neural plate, creating the **neural groove**. The edges, or **neural folds**, gradually move together. They first contact one another midway along the axis of the neural plate, near the end of the third week.

Where the neural folds meet, they fuse to form a cylindrical **neural tube** that loses its connection with the superficial ectoderm. The process of neural tube formation is called **neurulation**; it is completed in less than a week. The formation of the axial skeleton and that of the musculature around the developing neural tube were described in Embryology Summaries 7 and 9.

Cells at the tips of the neural folds do not participate in neural tube formation. These cells of the **neural crest** at first remain between the dorsal surface of the neural tube and the ectoderm, but they later migrate to other locations. The neural tube becomes the CNS. Axons from neurons within the neural tube and the axons of neural crest cells form the PNS.

23 DAYS

The first cells to appear in the mantle differentiate into neurons, while the last cells to arrive become astrocytes and oligodendrocytes. Further development of the CNS and PNS will be found in Embryology Summaries 11 and 12.

The neural tube increases in thickness as its epithelial lining undergoes repeated mitoses. By the middle of the fifth developmental week, there are three distinct layers. The **ependymal layer** lines the enclosed cavity, or **neurocoel**. The ependymal cells continue their mitotic activities, and daughter cells create the surrounding **mantle layer**. Axons from developing neurons form a superficial **marginal layer**.

EMBRYOLOGY SUMMARY 10: AN INTRODUCTION TO THE DEVELOPMENT OF THE NERVOUS SYSTEM

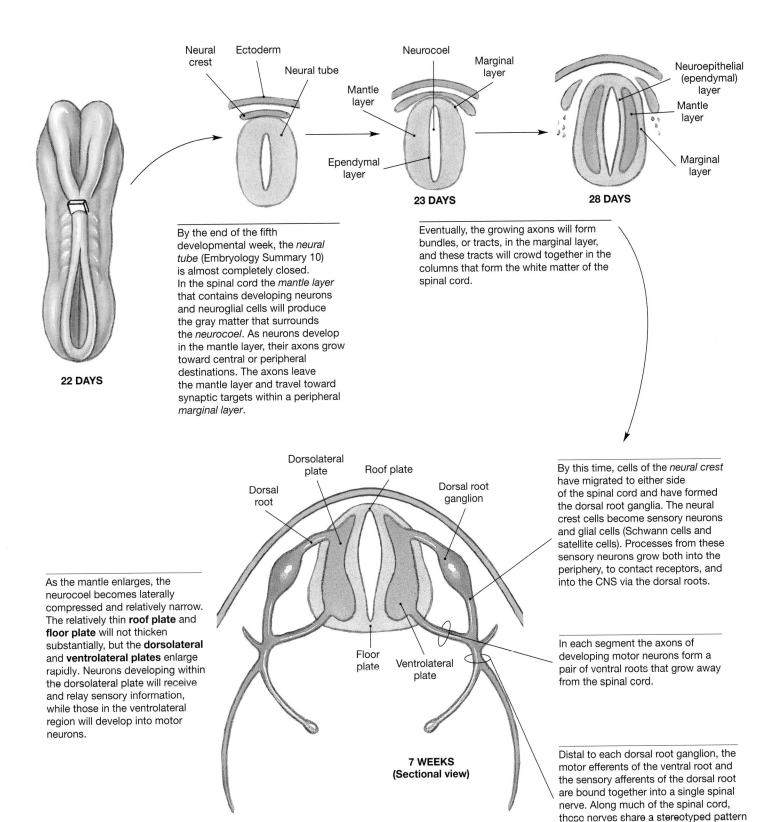

22 DAYS

Neural crest · Ectoderm · Neural tube · Mantle layer · Ependymal layer

23 DAYS

Neurocoel · Marginal layer

28 DAYS

Neuroepithelial (ependymal) layer · Mantle layer · Marginal layer

By the end of the fifth developmental week, the *neural tube* (Embryology Summary 10) is almost completely closed. In the spinal cord the *mantle layer* that contains developing neurons and neuroglial cells will produce the gray matter that surrounds the *neurocoel*. As neurons develop in the mantle layer, their axons grow toward central or peripheral destinations. The axons leave the mantle layer and travel toward synaptic targets within a peripheral *marginal layer*.

Eventually, the growing axons will form bundles, or tracts, in the marginal layer, and these tracts will crowd together in the columns that form the white matter of the spinal cord.

Dorsolateral plate · **Roof plate** · Dorsal root · Dorsal root ganglion · Floor plate · Ventrolateral plate

7 WEEKS (Sectional view)

As the mantle enlarges, the neurocoel becomes laterally compressed and relatively narrow. The relatively thin **roof plate** and **floor plate** will not thicken substantially, but the **dorsolateral** and **ventrolateral plates** enlarge rapidly. Neurons developing within the dorsolateral plate will receive and relay sensory information, while those in the ventrolateral region will develop into motor neurons.

By this time, cells of the *neural crest* have migrated to either side of the spinal cord and have formed the dorsal root ganglia. The neural crest cells become sensory neurons and glial cells (Schwann cells and satellite cells). Processes from these sensory neurons grow both into the periphery, to contact receptors, and into the CNS via the dorsal roots.

In each segment the axons of developing motor neurons form a pair of ventral roots that grow away from the spinal cord.

Distal to each dorsal root ganglion, the motor efferents of the ventral root and the sensory afferents of the dorsal root are bound together into a single spinal nerve. Along much of the spinal cord, these nerves share a stereotyped pattern of peripheral branches, and the pattern accounts for the distribution of dermatomes.

EMBRYOLOGY SUMMARY 11: THE DEVELOPMENT OF THE SPINAL CORD AND SPINAL NERVES—PART A

Neural crest cells aggregate to form autonomic ganglia near the vertebral column and in peripheral organs. Migrating neural crest cells contribute to the formation of teeth and form the laryngeal cartilages, melanocytes of the skin, the skull, connective tissues around the eye, the intrinsic muscles of the eye, Schwann cells, amphicytes, and the adrenal medullae.

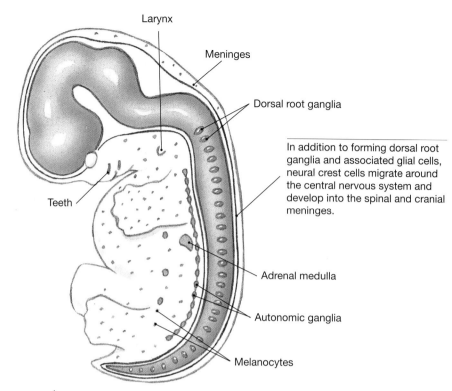

Larynx

Meninges

Dorsal root ganglia

Teeth

In addition to forming dorsal root ganglia and associated glial cells, neural crest cells migrate around the central nervous system and develop into the spinal and cranial meninges.

Adrenal medulla

Autonomic ganglia

Melanocytes

7 WEEKS
(Distribution of neural crest cells)

Several spinal nerves innervate each developing limb. When embryonic muscle cells migrate away from the myotome, the nerves grow right along with them. If a large muscle in the adult is derived from several myotomal blocks, connective tissue partitions will often mark the original boundaries, and the innervation will always involve more than one spinal nerve.

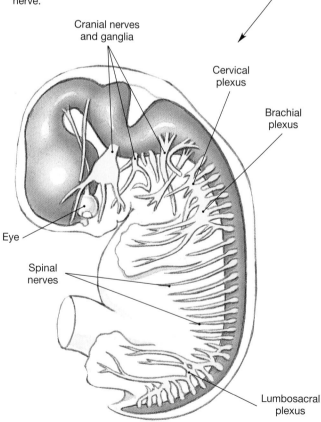

Cranial nerves and ganglia

Cervical plexus

Brachial plexus

Eye

Spinal nerves

Lumbosacral plexus

7 WEEKS
(Peripheral nerve distribution)

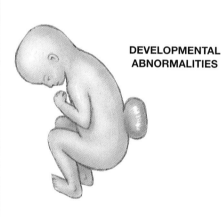

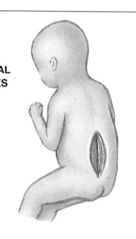

DEVELOPMENTAL ABNORMALITIES

Spina bifida

Neural tube defect

Spina bifida (BI-fi-da) results when the developing vertebral laminae fail to unite due to abnormal neural tube formation at that site. The neural arch is incomplete, and the meninges bulge outward beneath the skin of the back. The extent of the abnormality determines the severity of the defects. In mild cases, the condition may pass unnoticed; extreme cases involve much of the length of the vertebral column.

A **neural tube defect (NTD)** is a condition that is secondary to a developmental error in the formation of the spinal cord. Instead of forming a hollow tube, a portion of the spinal cord develops as a broad plate. This is often associated with spina bifida. Neural tube defects affect roughly one individual in 1000; prenatal testing can detect the existence of these defects with an 80–85% success rate.

EMBRYOLOGY SUMMARY 11: THE DEVELOPMENT OF THE SPINAL CORD AND SPINAL NERVES—PART B

Before proceeding, briefly review the summaries of skull formation (Embryology Summary 6) and vertebral column development (Embryology Summary 11).

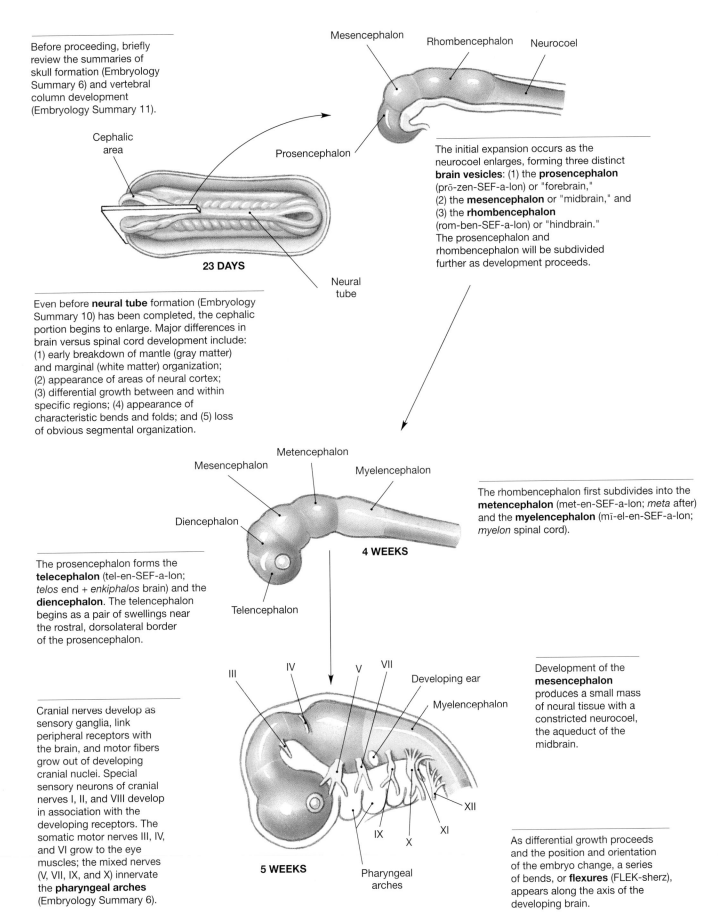

Mesencephalon Rhombencephalon Neurocoel

Cephalic area

Prosencephalon

23 DAYS

Neural tube

The initial expansion occurs as the neurocoel enlarges, forming three distinct **brain vesicles**: (1) the **prosencephalon** (prō-zen-SEF-a-lon) or "forebrain," (2) the **mesencephalon** or "midbrain," and (3) the **rhombencephalon** (rom-ben-SEF-a-lon) or "hindbrain." The prosencephalon and rhombencephalon will be subdivided further as development proceeds.

Even before **neural tube** formation (Embryology Summary 10) has been completed, the cephalic portion begins to enlarge. Major differences in brain versus spinal cord development include: (1) early breakdown of mantle (gray matter) and marginal (white matter) organization; (2) appearance of areas of neural cortex; (3) differential growth between and within specific regions; (4) appearance of characteristic bends and folds; and (5) loss of obvious segmental organization.

Mesencephalon Metencephalon Myelencephalon

Diencephalon

Telencephalon

4 WEEKS

The rhombencephalon first subdivides into the **metencephalon** (met-en-SEF-a-lon; *meta* after) and the **myelencephalon** (mī-el-en-SEF-a-lon; *myelon* spinal cord).

The prosencephalon forms the **telecephalon** (tel-en-SEF-a-lon; *telos* end + *enkiphalos* brain) and the **diencephalon**. The telencephalon begins as a pair of swellings near the rostral, dorsolateral border of the prosencephalon.

Cranial nerves develop as sensory ganglia, link peripheral receptors with the brain, and motor fibers grow out of developing cranial nuclei. Special sensory neurons of cranial nerves I, II, and VIII develop in association with the developing receptors. The somatic motor nerves III, IV, and VI grow to the eye muscles; the mixed nerves (V, VII, IX, and X) innervate the **pharyngeal arches** (Embryology Summary 6).

III IV V VII

Developing ear

Myelencephalon

5 WEEKS

Pharyngeal arches

IX X XI XII

Development of the **mesencephalon** produces a small mass of neural tissue with a constricted neurocoel, the aqueduct of the midbrain.

As differential growth proceeds and the position and orientation of the embryo change, a series of bends, or **flexures** (FLEK-sherz), appears along the axis of the developing brain.

EMBRYOLOGY SUMMARY 12: THE DEVELOPMENT OF THE BRAIN AND CRANIAL NERVES—PART A

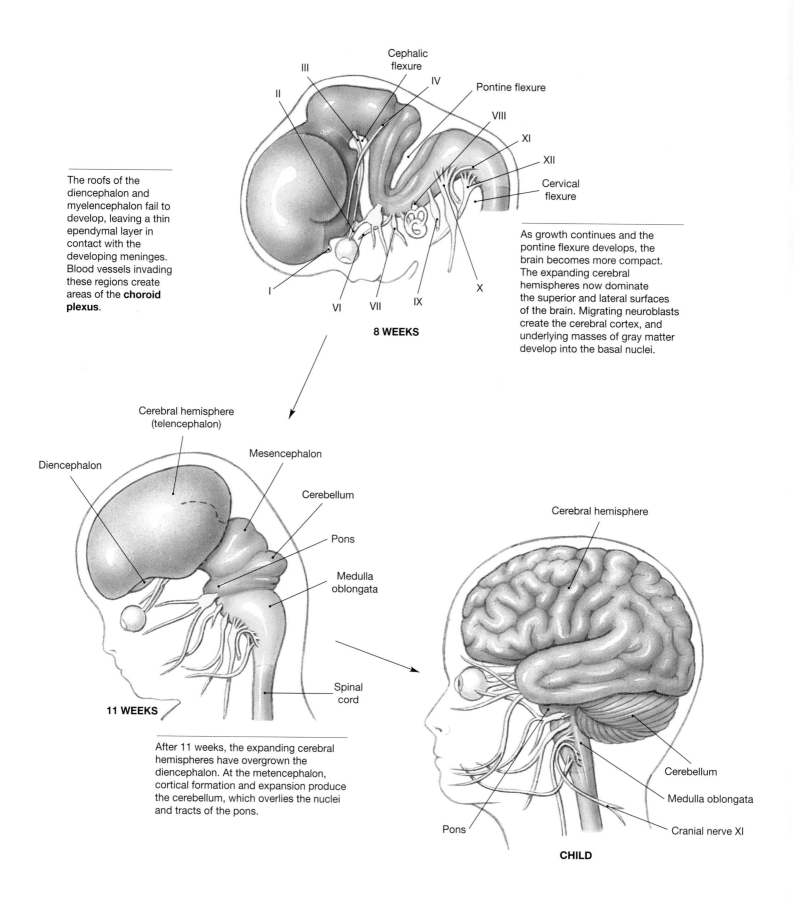

The roofs of the diencephalon and myelencephalon fail to develop, leaving a thin ependymal layer in contact with the developing meninges. Blood vessels invading these regions create areas of the **choroid plexus**.

8 WEEKS

As growth continues and the pontine flexure develops, the brain becomes more compact. The expanding cerebral hemispheres now dominate the superior and lateral surfaces of the brain. Migrating neuroblasts create the cerebral cortex, and underlying masses of gray matter develop into the basal nuclei.

11 WEEKS

After 11 weeks, the expanding cerebral hemispheres have overgrown the diencephalon. At the metencephalon, cortical formation and expansion produce the cerebellum, which overlies the nuclei and tracts of the pons.

CHILD

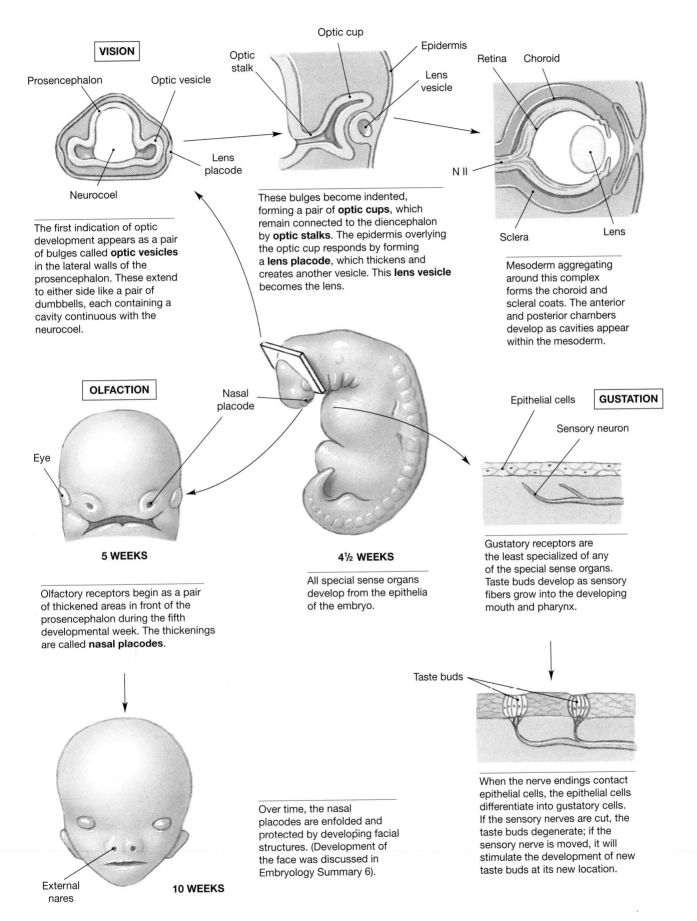

VISION

Prosencephalon Optic vesicle

Neurocoel

The first indication of optic development appears as a pair of bulges called **optic vesicles** in the lateral walls of the prosencephalon. These extend to either side like a pair of dumbbells, each containing a cavity continuous with the neurocoel.

Optic cup

Optic stalk Epidermis

Lens vesicle

Lens placode

These bulges become indented, forming a pair of **optic cups**, which remain connected to the diencephalon by **optic stalks**. The epidermis overlying the optic cup responds by forming a **lens placode**, which thickens and creates another vesicle. This **lens vesicle** becomes the lens.

Retina Choroid

N II

Sclera Lens

Mesoderm aggregating around this complex forms the choroid and scleral coats. The anterior and posterior chambers develop as cavities appear within the mesoderm.

OLFACTION

Nasal placode

Eye

5 WEEKS

Olfactory receptors begin as a pair of thickened areas in front of the prosencephalon during the fifth developmental week. The thickenings are called **nasal placodes**.

4½ WEEKS

All special sense organs develop from the epithelia of the embryo.

Epithelial cells **GUSTATION**

Sensory neuron

Gustatory receptors are the least specialized of any of the special sense organs. Taste buds develop as sensory fibers grow into the developing mouth and pharynx.

External nares

10 WEEKS

Over time, the nasal placodes are enfolded and protected by developing facial structures. (Development of the face was discussed in Embryology Summary 6).

Taste buds

When the nerve endings contact epithelial cells, the epithelial cells differentiate into gustatory cells. If the sensory nerves are cut, the taste buds degenerate; if the sensory nerve is moved, it will stimulate the development of new taste buds at its new location.

EMBRYOLOGY SUMMARY 13: **THE DEVELOPMENT OF SPECIAL SENSE ORGANS—PART A**

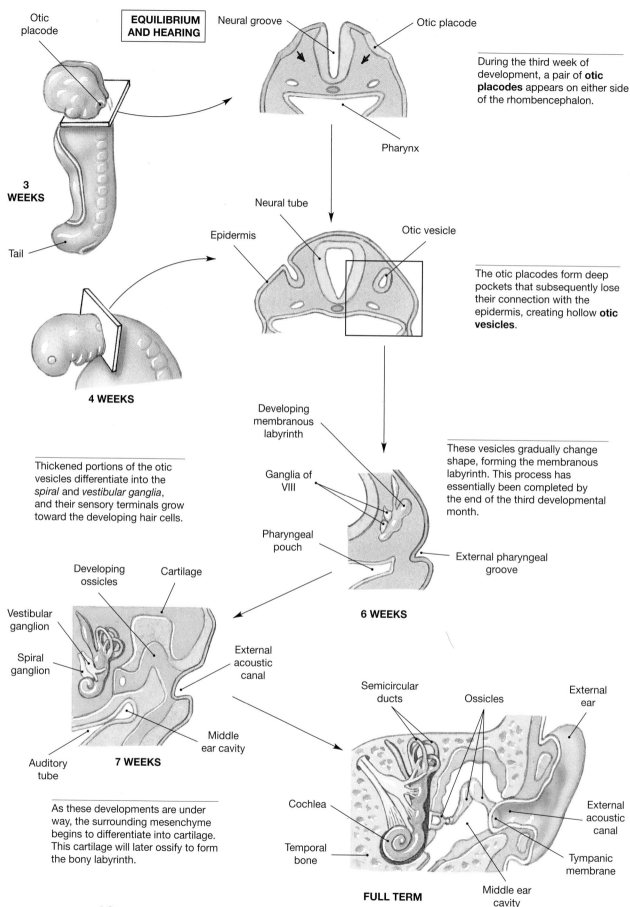

EQUILIBRIUM AND HEARING

During the third week of development, a pair of **otic placodes** appears on either side of the rhombencephalon.

The otic placodes form deep pockets that subsequently lose their connection with the epidermis, creating hollow **otic vesicles**.

These vesicles gradually change shape, forming the membranous labyrinth. This process has essentially been completed by the end of the third developmental month.

Thickened portions of the otic vesicles differentiate into the *spiral* and *vestibular ganglia*, and their sensory terminals grow toward the developing hair cells.

As these developments are under way, the surrounding mesenchyme begins to differentiate into cartilage. This cartilage will later ossify to form the bony labyrinth.

3 WEEKS

4 WEEKS

6 WEEKS

7 WEEKS

FULL TERM

EMBRYOLOGY SUMMARY 13: THE DEVELOPMENT OF SPECIAL SENSE ORGANS—PART B

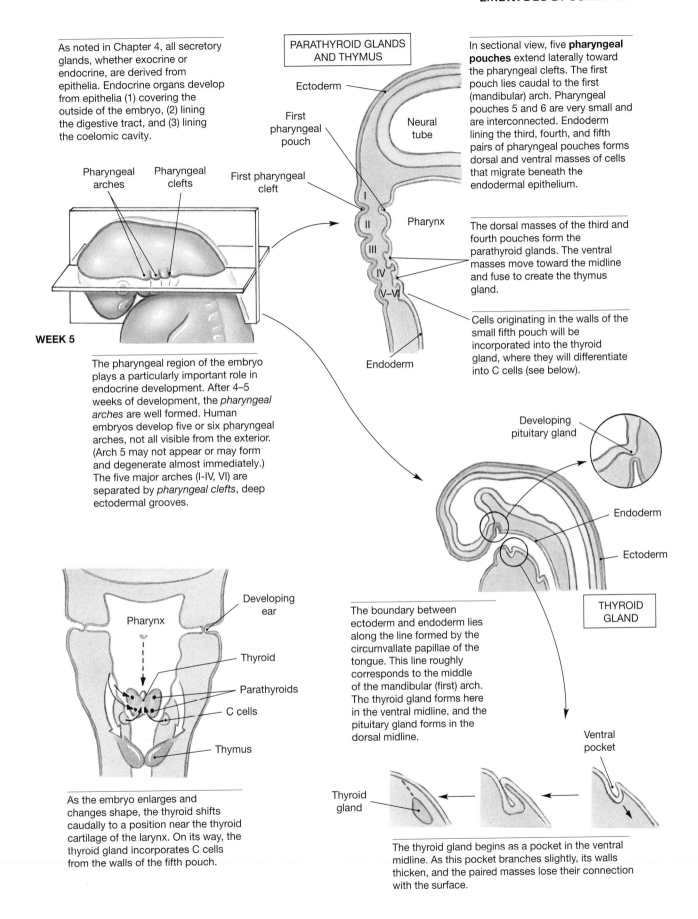

As noted in Chapter 4, all secretory glands, whether exocrine or endocrine, are derived from epithelia. Endocrine organs develop from epithelia (1) covering the outside of the embryo, (2) lining the digestive tract, and (3) lining the coelomic cavity.

PARATHYROID GLANDS AND THYMUS

Ectoderm

First pharyngeal pouch

First pharyngeal cleft

Pharyngeal arches

Pharyngeal clefts

WEEK 5

Neural tube

Pharynx

Endoderm

In sectional view, five **pharyngeal pouches** extend laterally toward the pharyngeal clefts. The first pouch lies caudal to the first (mandibular) arch. Pharyngeal pouches 5 and 6 are very small and are interconnected. Endoderm lining the third, fourth, and fifth pairs of pharyngeal pouches forms dorsal and ventral masses of cells that migrate beneath the endodermal epithelium.

The dorsal masses of the third and fourth pouches form the parathyroid glands. The ventral masses move toward the midline and fuse to create the thymus gland.

Cells originating in the walls of the small fifth pouch will be incorporated into the thyroid gland, where they will differentiate into C cells (see below).

The pharyngeal region of the embryo plays a particularly important role in endocrine development. After 4–5 weeks of development, the *pharyngeal arches* are well formed. Human embryos develop five or six pharyngeal arches, not all visible from the exterior. (Arch 5 may not appear or may form and degenerate almost immediately.) The five major arches (I-IV, VI) are separated by *pharyngeal clefts*, deep ectodermal grooves.

Developing pituitary gland

Endoderm

Ectoderm

THYROID GLAND

Pharynx

Developing ear

Thyroid

Parathyroids

C cells

Thymus

As the embryo enlarges and changes shape, the thyroid shifts caudally to a position near the thyroid cartilage of the larynx. On its way, the thyroid gland incorporates C cells from the walls of the fifth pouch.

The boundary between ectoderm and endoderm lies along the line formed by the circumvallate papillae of the tongue. This line roughly corresponds to the middle of the mandibular (first) arch. The thyroid gland forms here in the ventral midline, and the pituitary gland forms in the dorsal midline.

Ventral pocket

Thyroid gland

The thyroid gland begins as a pocket in the ventral midline. As this pocket branches slightly, its walls thicken, and the paired masses lose their connection with the surface.

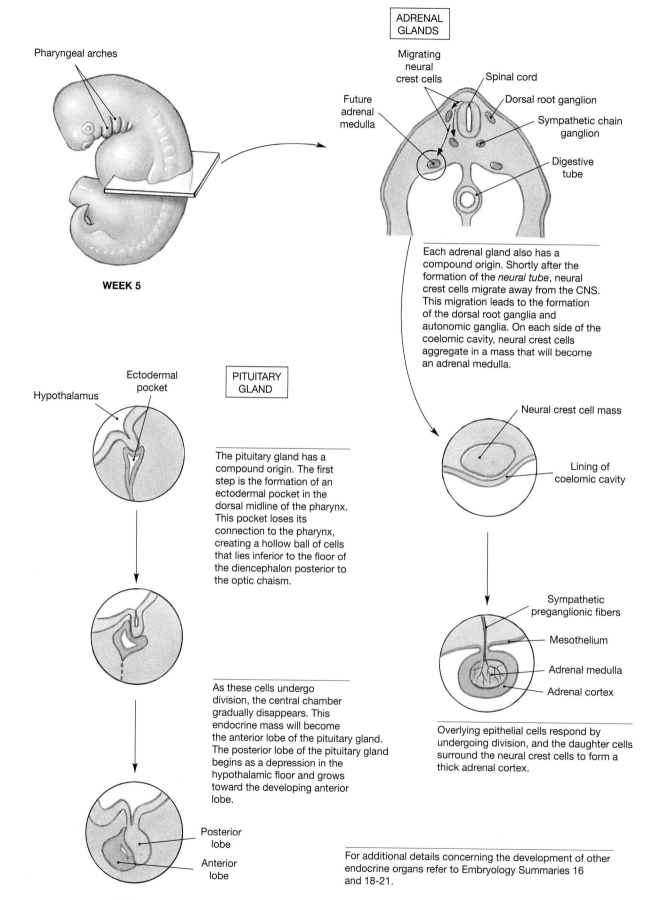

Pharyngeal arches

WEEK 5

ADRENAL GLANDS

Migrating neural crest cells

Spinal cord

Future adrenal medulla

Dorsal root ganglion

Sympathetic chain ganglion

Digestive tube

Each adrenal gland also has a compound origin. Shortly after the formation of the *neural tube*, neural crest cells migrate away from the CNS. This migration leads to the formation of the dorsal root ganglia and autonomic ganglia. On each side of the coelomic cavity, neural crest cells aggregate in a mass that will become an adrenal medulla.

Neural crest cell mass

Lining of coelomic cavity

Sympathetic preganglionic fibers

Mesothelium

Adrenal medulla

Adrenal cortex

Overlying epithelial cells respond by undergoing division, and the daughter cells surround the neural crest cells to form a thick adrenal cortex.

Hypothalamus

Ectodermal pocket

PITUITARY GLAND

The pituitary gland has a compound origin. The first step is the formation of an ectodermal pocket in the dorsal midline of the pharynx. This pocket loses its connection to the pharynx, creating a hollow ball of cells that lies inferior to the floor of the diencephalon posterior to the optic chaism.

As these cells undergo division, the central chamber gradually disappears. This endocrine mass will become the anterior lobe of the pituitary gland. The posterior lobe of the pituitary gland begins as a depression in the hypothalamic floor and grows toward the developing anterior lobe.

Posterior lobe

Anterior lobe

For additional details concerning the development of other endocrine organs refer to Embryology Summaries 16 and 18-21.

EMBRYOLOGY SUMMARY **14:** THE DEVELOPMENT OF THE ENDOCRINE SYSTEM—*PART B*

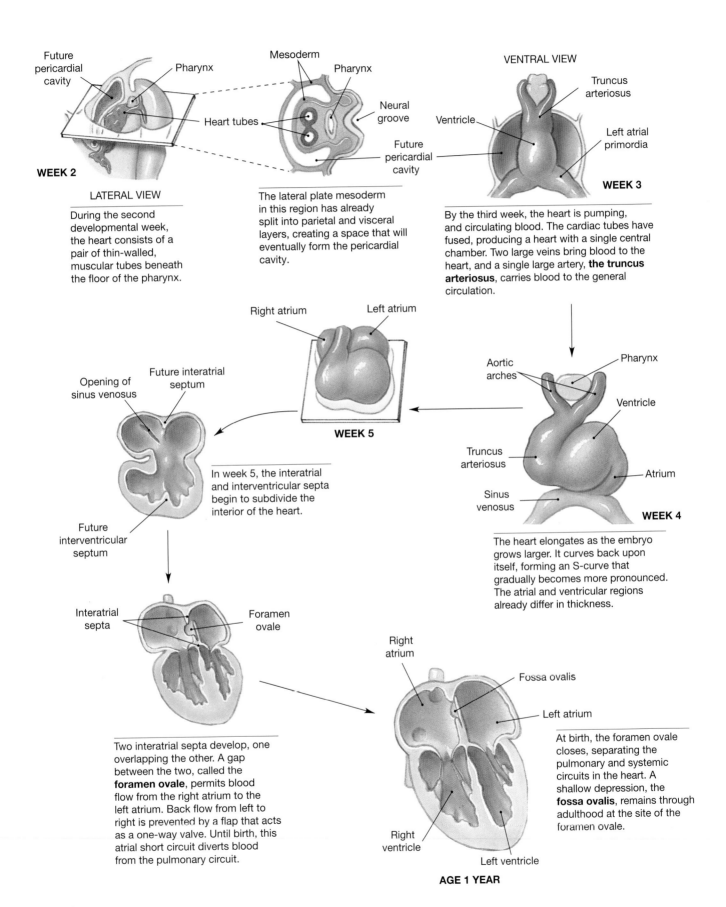

Future pericardial cavity

Pharynx

Heart tubes

WEEK 2

LATERAL VIEW

During the second developmental week, the heart consists of a pair of thin-walled, muscular tubes beneath the floor of the pharynx.

Mesoderm

Pharynx

Neural groove

Future pericardial cavity

The lateral plate mesoderm in this region has already split into parietal and visceral layers, creating a space that will eventually form the pericardial cavity.

VENTRAL VIEW

Truncus arteriosus

Ventricle

Left atrial primordia

WEEK 3

By the third week, the heart is pumping, and circulating blood. The cardiac tubes have fused, producing a heart with a single central chamber. Two large veins bring blood to the heart, and a single large artery, **the truncus arteriosus**, carries blood to the general circulation.

Right atrium

Left atrium

WEEK 5

Opening of sinus venosus

Future interatrial septum

Future interventricular septum

In week 5, the interatrial and interventricular septa begin to subdivide the interior of the heart.

Aortic arches

Pharynx

Ventricle

Truncus arteriosus

Sinus venosus

Atrium

WEEK 4

The heart elongates as the embryo grows larger. It curves back upon itself, forming an S-curve that gradually becomes more pronounced. The atrial and ventricular regions already differ in thickness.

Interatrial septa

Foramen ovale

Two interatrial septa develop, one overlapping the other. A gap between the two, called the **foramen ovale**, permits blood flow from the right atrium to the left atrium. Back flow from left to right is prevented by a flap that acts as a one-way valve. Until birth, this atrial short circuit diverts blood from the pulmonary circuit.

Right atrium

Fossa ovalis

Left atrium

Right ventricle

Left ventricle

AGE 1 YEAR

At birth, the foramen ovale closes, separating the pulmonary and systemic circuits in the heart. A shallow depression, the **fossa ovalis**, remains through adulthood at the site of the foramen ovale.

EMBRYOLOGY SUMMARY 15: THE DEVELOPMENT OF THE HEART

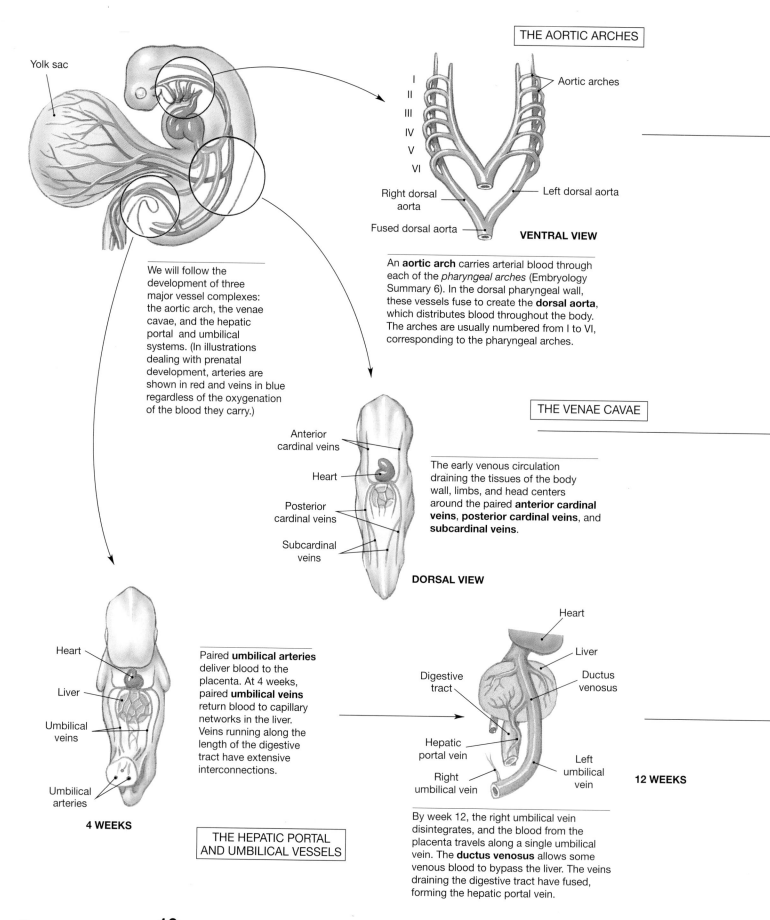

Yolk sac

THE AORTIC ARCHES

I
II
III
IV
V
VI

Aortic arches

Right dorsal aorta

Left dorsal aorta

Fused dorsal aorta

VENTRAL VIEW

An **aortic arch** carries arterial blood through each of the *pharyngeal arches* (Embryology Summary 6). In the dorsal pharyngeal wall, these vessels fuse to create the **dorsal aorta**, which distributes blood throughout the body. The arches are usually numbered from I to VI, corresponding to the pharyngeal arches.

We will follow the development of three major vessel complexes: the aortic arch, the venae cavae, and the hepatic portal and umbilical systems. (In illustrations dealing with prenatal development, arteries are shown in red and veins in blue regardless of the oxygenation of the blood they carry.)

THE VENAE CAVAE

Anterior cardinal veins

Heart

Posterior cardinal veins

Subcardinal veins

DORSAL VIEW

The early venous circulation draining the tissues of the body wall, limbs, and head centers around the paired **anterior cardinal veins**, **posterior cardinal veins**, and **subcardinal veins**.

Heart

Liver

Umbilical veins

Umbilical arteries

4 WEEKS

Paired **umbilical arteries** deliver blood to the placenta. At 4 weeks, paired **umbilical veins** return blood to capillary networks in the liver. Veins running along the length of the digestive tract have extensive interconnections.

THE HEPATIC PORTAL AND UMBILICAL VESSELS

Heart

Digestive tract

Liver

Ductus venosus

Hepatic portal vein

Right umbilical vein

Left umbilical vein

12 WEEKS

By week 12, the right umbilical vein disintegrates, and the blood from the placenta travels along a single umbilical vein. The **ductus venosus** allows some venous blood to bypass the liver. The veins draining the digestive tract have fused, forming the hepatic portal vein.

EMBRYOLOGY SUMMARY **16:** THE DEVELOPMENT OF THE CARDIOVASCULAR SYSTEM

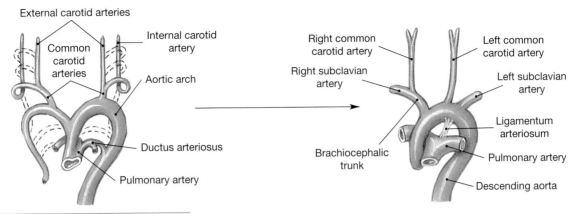

As development proceeds, some of these arches disintegrate. The **ductus arteriosus** provides an external short-circuit between the pulmonary and systemic circuits. Most of the blood entering the right atrium bypasses the lungs, passing instead through the ductus arteriosus or through the **foramen ovale** in the heart.

The left half of arch IV ultimately becomes the aortic arch, which carries blood away from the left ventricle.

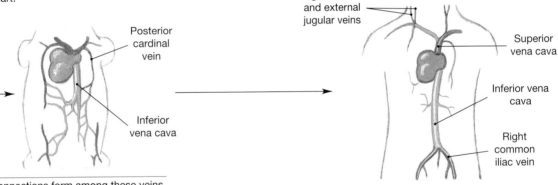

Interconnections form among these veins, and a combination of fusion and disintegration produces more-direct, larger-diameter connections to the right atrium.

This process continues, ultimately producing the superior and inferior venae cavae.

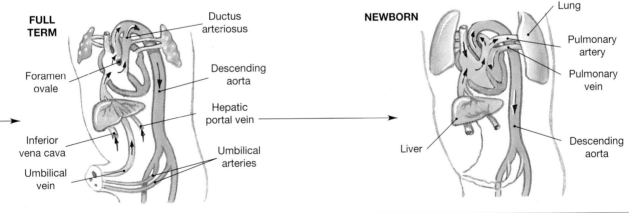

Shortly before birth, blood returning from the placenta travels through the liver in the ductus venosus to reach the inferior vena cava. Much of the blood delivered by the venae cavae bypasses the lungs by traveling through the foramen ovale and the ductus arteriosus.

At birth, pressures drop in the pleural cavities as the chest expands and the infant takes its first breath. The pulmonary vessels dilate, and blood flow to the lungs increases. Pressure falls in the right atrium, and the higher left atrial pressures close the valve that guards the foramen ovale. Smooth muscles contract the ductus arteriosus, which ultimately converts to the **ligamentum arteriosum**, a fibrous strand.

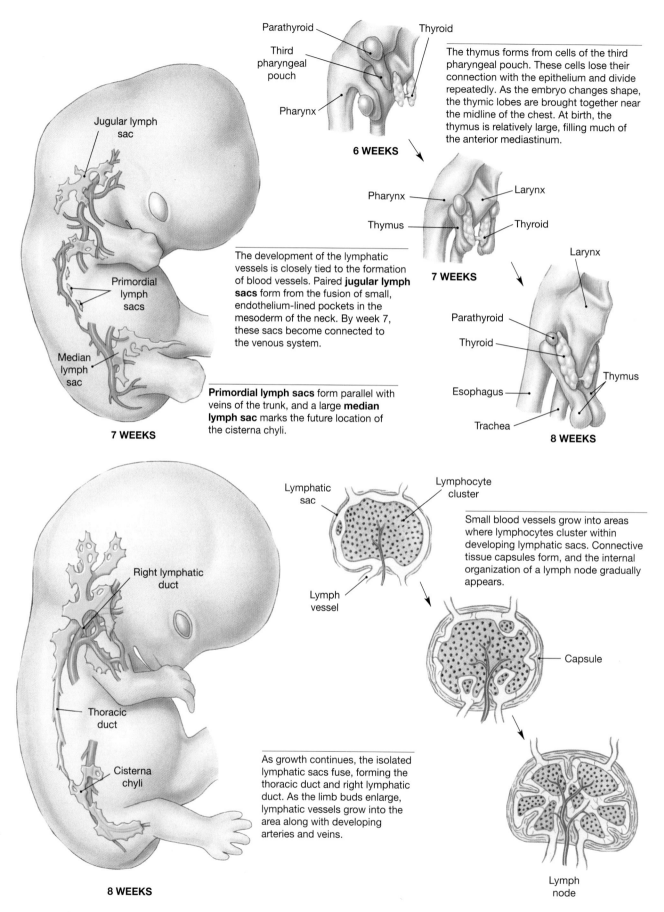

6 WEEKS

The thymus forms from cells of the third pharyngeal pouch. These cells lose their connection with the epithelium and divide repeatedly. As the embryo changes shape, the thymic lobes are brought together near the midline of the chest. At birth, the thymus is relatively large, filling much of the anterior mediastinum.

7 WEEKS

The development of the lymphatic vessels is closely tied to the formation of blood vessels. Paired **jugular lymph sacs** form from the fusion of small, endothelium-lined pockets in the mesoderm of the neck. By week 7, these sacs become connected to the venous system.

Primordial lymph sacs form parallel with veins of the trunk, and a large **median lymph sac** marks the future location of the cisterna chyli.

7 WEEKS

8 WEEKS

Small blood vessels grow into areas where lymphocytes cluster within developing lymphatic sacs. Connective tissue capsules form, and the internal organization of a lymph node gradually appears.

As growth continues, the isolated lymphatic sacs fuse, forming the thoracic duct and right lymphatic duct. As the limb buds enlarge, lymphatic vessels grow into the area along with developing arteries and veins.

8 WEEKS

EMBRYOLOGY SUMMARY **17**: THE DEVELOPMENT OF THE LYMPHATIC SYSTEM

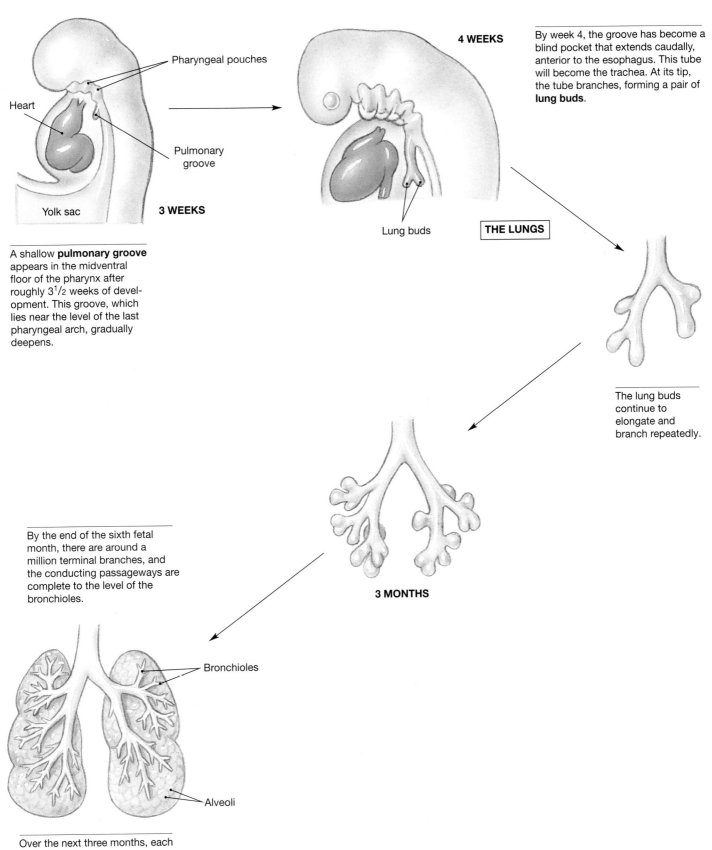

Pharyngeal pouches

Heart

Pulmonary groove

Yolk sac

3 WEEKS

A shallow **pulmonary groove** appears in the midventral floor of the pharynx after roughly 3¹/₂ weeks of development. This groove, which lies near the level of the last pharyngeal arch, gradually deepens.

4 WEEKS

By week 4, the groove has become a blind pocket that extends caudally, anterior to the esophagus. This tube will become the trachea. At its tip, the tube branches, forming a pair of **lung buds**.

Lung buds

THE LUNGS

The lung buds continue to elongate and branch repeatedly.

By the end of the sixth fetal month, there are around a million terminal branches, and the conducting passageways are complete to the level of the bronchioles.

3 MONTHS

Bronchioles

Alveoli

Over the next three months, each of the bronchioles gives rise to several hundred alveoli. This process continues for a variable period after birth.

EMBRYOLOGY SUMMARY 18: **THE DEVELOPMENT OF THE RESPIRATORY SYSTEM—*PART A***

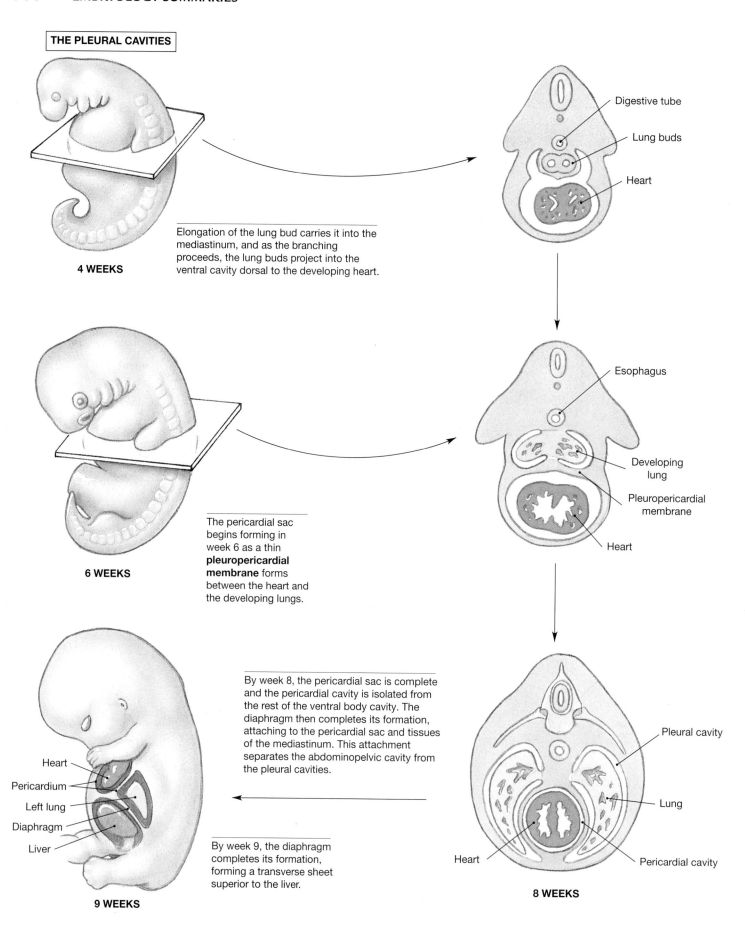

THE PLEURAL CAVITIES

Elongation of the lung bud carries it into the mediastinum, and as the branching proceeds, the lung buds project into the ventral cavity dorsal to the developing heart.

4 WEEKS

Digestive tube

Lung buds

Heart

The pericardial sac begins forming in week 6 as a thin **pleuropericardial membrane** forms between the heart and the developing lungs.

6 WEEKS

Esophagus

Developing lung

Pleuropericardial membrane

Heart

By week 8, the pericardial sac is complete and the pericardial cavity is isolated from the rest of the ventral body cavity. The diaphragm then completes its formation, attaching to the pericardial sac and tissues of the mediastinum. This attachment separates the abdominopelvic cavity from the pleural cavities.

Heart

Pericardium

Left lung

Diaphragm

Liver

By week 9, the diaphragm completes its formation, forming a transverse sheet superior to the liver.

9 WEEKS

Pleural cavity

Lung

Pericardial cavity

Heart

8 WEEKS

EMBRYOLOGY SUMMARY **18:** THE DEVELOPMENT OF THE RESPIRATORY SYSTEM—*PART B*

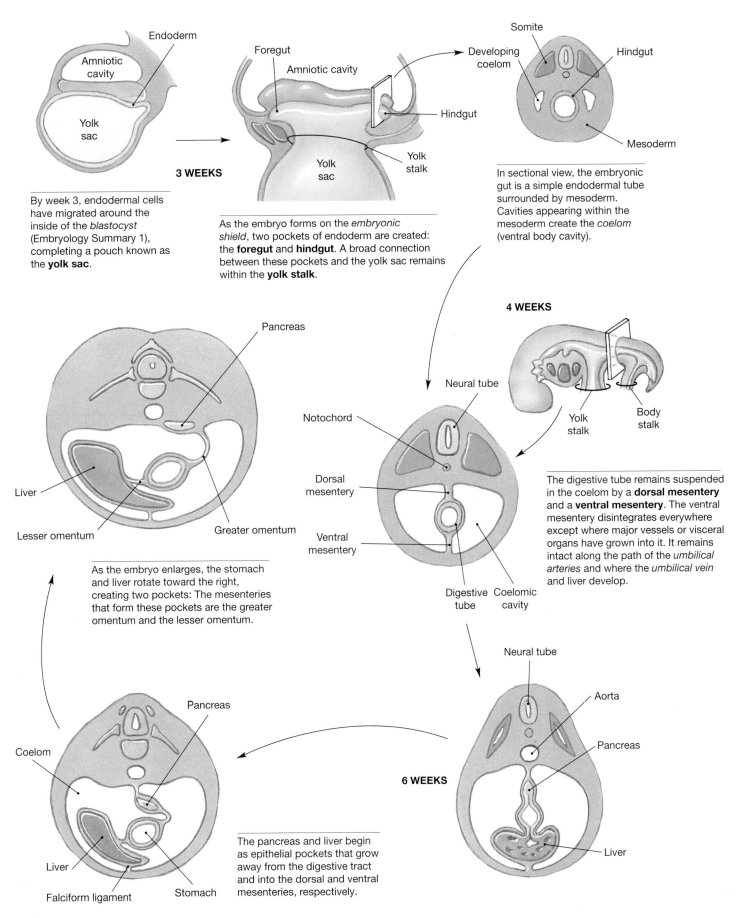

3 WEEKS

By week 3, endodermal cells have migrated around the inside of the *blastocyst* (Embryology Summary 1), completing a pouch known as the **yolk sac**.

As the embryo forms on the *embryonic shield*, two pockets of endoderm are created: the **foregut** and **hindgut**. A broad connection between these pockets and the yolk sac remains within the **yolk stalk**.

In sectional view, the embryonic gut is a simple endodermal tube surrounded by mesoderm. Cavities appearing within the mesoderm create the *coelom* (ventral body cavity).

4 WEEKS

The digestive tube remains suspended in the coelom by a **dorsal mesentery** and a **ventral mesentery**. The ventral mesentery disintegrates everywhere except where major vessels or visceral organs have grown into it. It remains intact along the path of the *umbilical arteries* and where the *umbilical vein* and liver develop.

As the embryo enlarges, the stomach and liver rotate toward the right, creating two pockets: The mesenteries that form these pockets are the greater omentum and the lesser omentum.

The pancreas and liver begin as epithelial pockets that grow away from the digestive tract and into the dorsal and ventral mesenteries, respectively.

6 WEEKS

EMBRYOLOGY SUMMARY 19: THE DEVELOPMENT OF THE DIGESTIVE SYSTEM—PART A

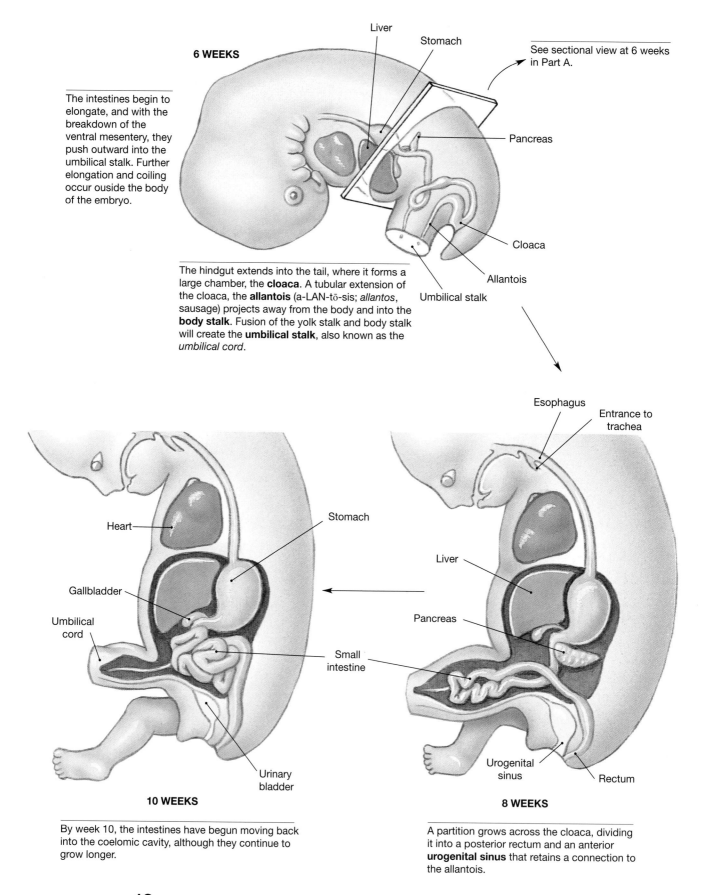

6 WEEKS

Liver

Stomach

See sectional view at 6 weeks in Part A.

The intestines begin to elongate, and with the breakdown of the ventral mesentery, they push outward into the umbilical stalk. Further elongation and coiling occur ouside the body of the embryo.

Pancreas

Cloaca

Allantois

Umbilical stalk

The hindgut extends into the tail, where it forms a large chamber, the **cloaca**. A tubular extension of the cloaca, the **allantois** (a-LAN-tō-sis; *allantos*, sausage) projects away from the body and into the **body stalk**. Fusion of the yolk stalk and body stalk will create the **umbilical stalk**, also known as the *umbilical cord*.

Esophagus

Entrance to trachea

Heart

Stomach

Liver

Gallbladder

Pancreas

Umbilical cord

Small intestine

Urogenital sinus

Rectum

Urinary bladder

10 WEEKS

By week 10, the intestines have begun moving back into the coelomic cavity, although they continue to grow longer.

8 WEEKS

A partition grows across the cloaca, dividing it into a posterior rectum and an anterior **urogenital sinus** that retains a connection to the allantois.

EMBRYOLOGY SUMMARY 19: THE DEVELOPMENT OF THE DIGESTIVE SYSTEM—*PART B*

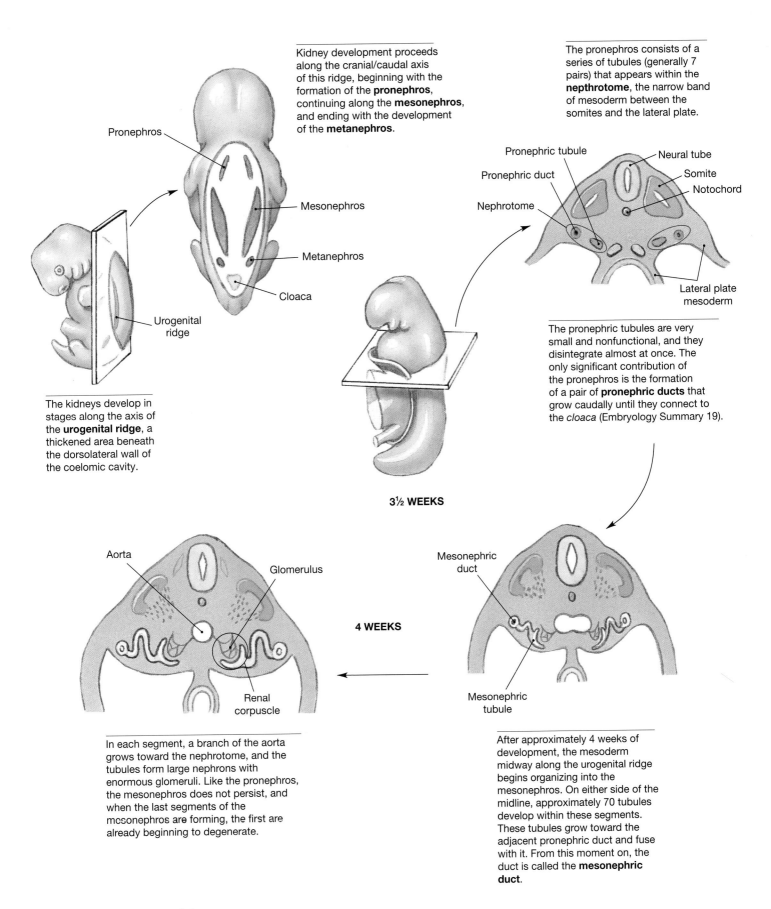

Kidney development proceeds along the cranial/caudal axis of this ridge, beginning with the formation of the **pronephros**, continuing along the **mesonephros**, and ending with the development of the **metanephros**.

The pronephros consists of a series of tubules (generally 7 pairs) that appears within the **nephrotome**, the narrow band of mesoderm between the somites and the lateral plate.

Pronephros

Mesonephros

Metanephros

Cloaca

Urogenital ridge

Pronephric tubule

Pronephric duct

Nephrotome

Neural tube

Somite

Notochord

Lateral plate mesoderm

The kidneys develop in stages along the axis of the **urogenital ridge**, a thickened area beneath the dorsolateral wall of the coelomic cavity.

The pronephric tubules are very small and nonfunctional, and they disintegrate almost at once. The only significant contribution of the pronephros is the formation of a pair of **pronephric ducts** that grow caudally until they connect to the *cloaca* (Embryology Summary 19).

3½ WEEKS

4 WEEKS

Aorta

Glomerulus

Renal corpuscle

Mesonephric duct

Mesonephric tubule

In each segment, a branch of the aorta grows toward the nephrotome, and the tubules form large nephrons with enormous glomeruli. Like the pronephros, the mesonephros does not persist, and when the last segments of the mesonephros are forming, the first are already beginning to degenerate.

After approximately 4 weeks of development, the mesoderm midway along the urogenital ridge begins organizing into the mesonephros. On either side of the midline, approximately 70 tubules develop within these segments. These tubules grow toward the adjacent pronephric duct and fuse with it. From this moment on, the duct is called the **mesonephric duct**.

EMBRYOLOGY SUMMARY 20: THE DEVELOPMENT OF THE URINARY SYSTEM—PART A

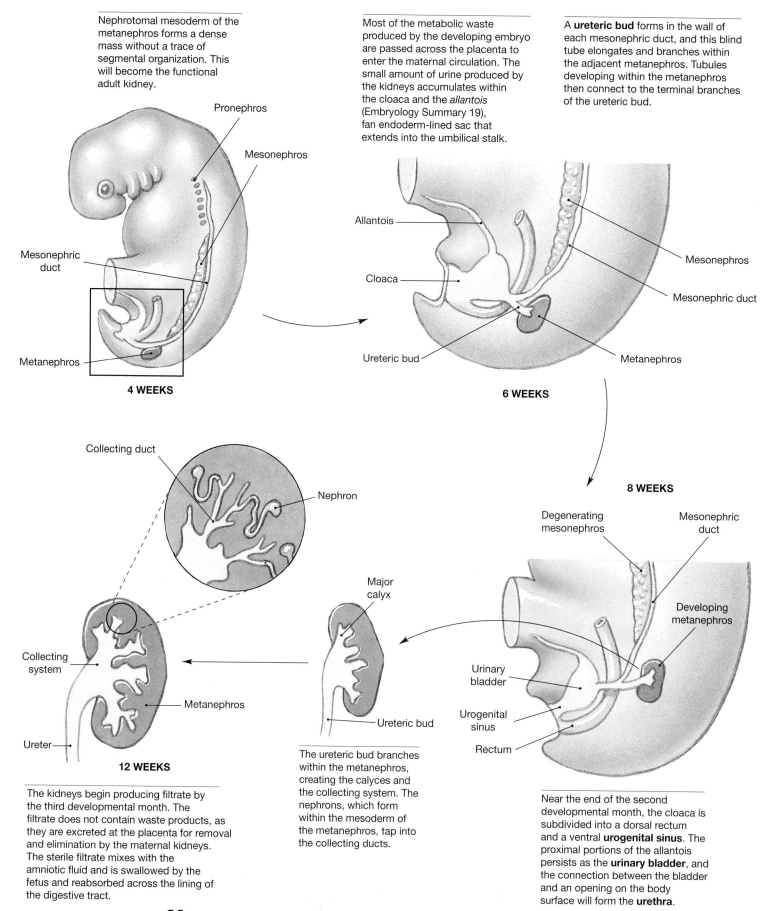

Nephrotomal mesoderm of the metanephros forms a dense mass without a trace of segmental organization. This will become the functional adult kidney.

Pronephros

Mesonephros

Mesonephric duct

Metanephros

4 WEEKS

Most of the metabolic waste produced by the developing embryo are passed across the placenta to enter the maternal circulation. The small amount of urine produced by the kidneys accumulates within the cloaca and the *allantois* (Embryology Summary 19), fan endoderm-lined sac that extends into the umbilical stalk.

Allantois

Cloaca

Ureteric bud

Mesonephros

Mesonephric duct

Metanephros

6 WEEKS

A **ureteric bud** forms in the wall of each mesonephric duct, and this blind tube elongates and branches within the adjacent metanephros. Tubules developing within the metanephros then connect to the terminal branches of the ureteric bud.

8 WEEKS

Collecting duct

Nephron

Collecting system

Metanephros

Ureter

12 WEEKS

The kidneys begin producing filtrate by the third developmental month. The filtrate does not contain waste products, as they are excreted at the placenta for removal and elimination by the maternal kidneys. The sterile filtrate mixes with the amniotic fluid and is swallowed by the fetus and reabsorbed across the lining of the digestive tract.

Major calyx

Ureteric bud

The ureteric bud branches within the metanephros, creating the calyces and the collecting system. The nephrons, which form within the mesoderm of the metanephros, tap into the collecting ducts.

Degenerating mesonephros

Mesonephric duct

Developing metanephros

Urinary bladder

Urogenital sinus

Rectum

Near the end of the second developmental month, the cloaca is subdivided into a dorsal rectum and a ventral **urogenital sinus**. The proximal portions of the allantois persists as the **urinary bladder**, and the connection between the bladder and an opening on the body surface will form the **urethra**.

EMBRYOLOGY SUMMARY 20: THE DEVELOPMENT OF THE URINARY SYSTEM—PART B

SEX-INDEPENDENT STAGES
(WEEKS 3–6)

DEVELOPMENT OF THE GONADS

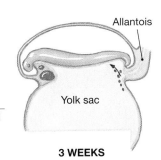

3 WEEKS

During the third week, endodermal cells migrate from the wall of the yolk sac near the allantois to the dorsal wall of the abdominal cavity. These primordial germ cells enter the **genital ridges** that parallel the mesonephros.

Each ridge has a thick epithelium continuous with columns of cells, the **primary sex cords**, that extend into the center (medulla) of the ridge. Anterior to each mesonephric duct, a duct forms that has no connection to the kidneys. This is the **paramesonephric (Müllerian) duct**; it extends along the genital ridge and continues toward the cloaca. At this gender-indifferent stage, male embryos cannot be distinguished from female embryos.

DEVELOPMENT OF DUCTS AND ACCESSORY ORGANS

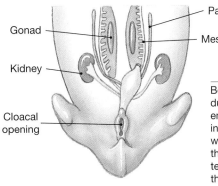

Both genders have mesonephric and paramesonephric ducts at this stage. Unless exposed to androgens, the embryo—regardless of its genetic gender—will develop into a female. (A genetic male who develops as a female will be sterile, however.) In a normal male embryo, cells in the core (medulla) of the genital ridge begin producing testosterone sometime after week 6. Testosterone triggers the changes in the duct system and external genitalia that are detailed in Part B.

DEVELOPMENT OF EXTERNAL GENITALIA

4 WEEKS

6 WEEKS

After 4 weeks of development, there are mesenchymal swellings called **cloacal folds** around the **cloacal membrane** (the cloaca does not open to the exterior). The **genital tubercle** forms the glans of the penis in males and the clitoris in females.

Two weeks later, the cloaca has been subdivided, separating the cloacal membrane into a posterior *anal membrane*, bounded by the *anal folds*, and an anterior **urogenital membrane**, bounded by the **urethral folds**. A prominent **genital swelling** forms lateral to each urethral fold.

EMBRYOLOGY SUMMARY **21**: THE DEVELOPMENT OF THE REPRODUCTIVE SYSTEM—*PART A*

DEVELOPMENT OF THE MALE REPRODUCTIVE SYSTEM

DEVELOPMENT OF THE TESTES

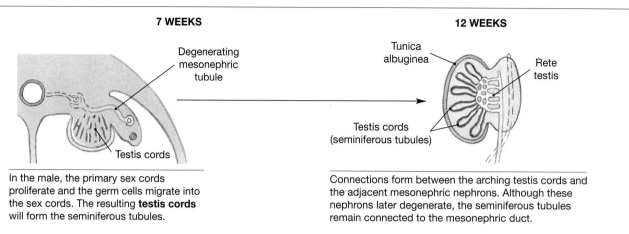

7 WEEKS

Degenerating mesonephric tubule

Testis cords

12 WEEKS

Tunica albuginea

Rete testis

Testis cords (seminiferous tubules)

In the male, the primary sex cords proliferate and the germ cells migrate into the sex cords. The resulting **testis cords** will form the seminiferous tubules.

Connections form between the arching testis cords and the adjacent mesonephric nephrons. Although these nephrons later degenerate, the seminiferous tubules remain connected to the mesonephric duct.

DEVELOPMENT OF MALE DUCTS AND ACCESSORY ORGANS

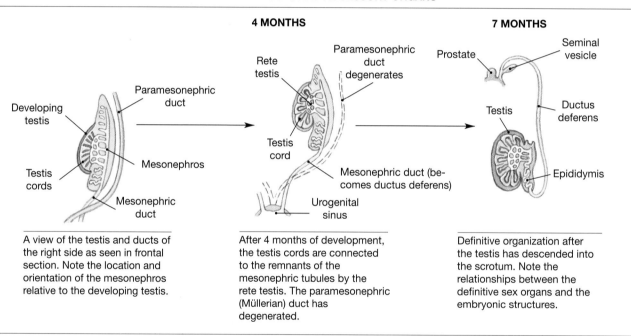

4 MONTHS

Developing testis

Paramesonephric duct

Testis cords

Mesonephros

Mesonephric duct

Rete testis

Paramesonephric duct degenerates

Testis cord

Mesonephric duct (becomes ductus deferens)

Urogenital sinus

7 MONTHS

Prostate

Seminal vesicle

Testis

Ductus deferens

Epididymis

A view of the testis and ducts of the right side as seen in frontal section. Note the location and orientation of the mesonephros relative to the developing testis.

After 4 months of development, the testis cords are connected to the remnants of the mesonephric tubules by the rete testis. The paramesonephric (Müllerian) duct has degenerated.

Definitive organization after the testis has descended into the scrotum. Note the relationships between the definitive sex organs and the embryonic structures.

DEVELOPMENT OF MALE EXTERNAL GENITALIA

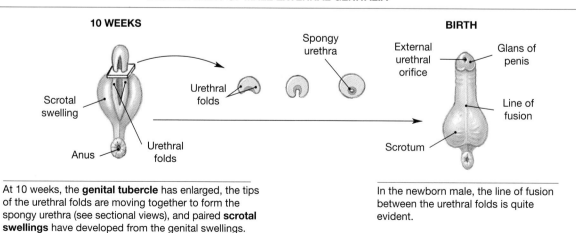

10 WEEKS

Scrotal swelling

Anus

Urethral folds

Urethral folds

Spongy urethra

BIRTH

External urethral orifice

Glans of penis

Line of fusion

Scrotum

At 10 weeks, the **genital tubercle** has enlarged, the tips of the urethral folds are moving together to form the spongy urethra (see sectional views), and paired **scrotal swellings** have developed from the genital swellings.

In the newborn male, the line of fusion between the urethral folds is quite evident.

DEVELOPMENT OF THE FEMALE REPRODUCTIVE SYSTEM

DEVELOPMENT OF THE OVARIES

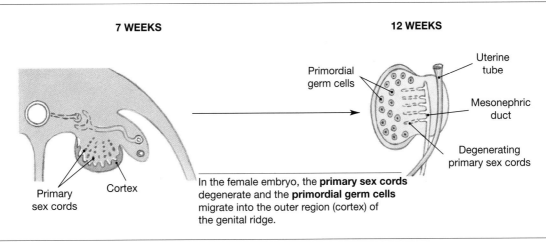

7 WEEKS

12 WEEKS

Primordial germ cells

Uterine tube

Mesonephric duct

Degenerating primary sex cords

Primary sex cords

Cortex

In the female embryo, the **primary sex cords** degenerate and the **primordial germ cells** migrate into the outer region (cortex) of the genital ridge.

DEVELOPMENT OF FEMALE DUCTS AND ACCESSORY ORGANS

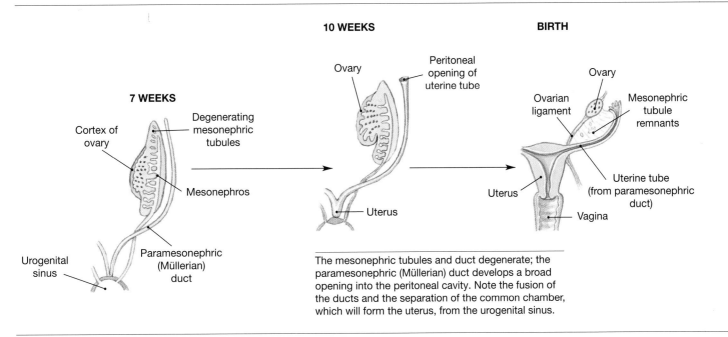

10 WEEKS

BIRTH

Ovary

Peritoneal opening of uterine tube

7 WEEKS

Cortex of ovary

Degenerating mesonephric tubules

Ovarian ligament

Ovary

Mesonephric tubule remnants

Mesonephros

Uterine tube (from paramesonephric duct)

Uterus

Urogenital sinus

Paramesonephric (Müllerian) duct

Uterus

Vagina

The mesonephric tubules and duct degenerate; the paramesonephric (Müllerian) duct develops a broad opening into the peritoneal cavity. Note the fusion of the ducts and the separation of the common chamber, which will form the uterus, from the urogenital sinus.

DEVELOPMENT OF FEMALE EXTERNAL GENITALIA

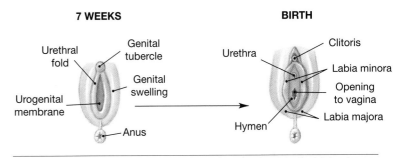

7 WEEKS

BIRTH

Urethral fold

Genital tubercle

Urethra

Clitoris

Labia minora

Genital swelling

Opening to vagina

Urogenital membrane

Labia majora

Anus

Hymen

COMPARISON OF MALE AND FEMALE EXTERNAL GENITALIA

Males	Females
Penis	Clitoris
Corpora cavernosa	Erectile tissue
Corpus spongiosum	Vestibular bulbs
Proximal shaft of penis	Labia minora
Spongy urethra	Vestibule
Bulbourethral glands	Greater vestibular glands
Scrotum	Labia majora

In the female, the urethral folds do not fuse; they develop into the labia minora. The genital swellings will form the labia majora. The genital tubercle develops into the clitoris. The urethra opens to the exterior immediately posterior to the clitoris. The hymen remains as an elaboration of the urogenital membrane.

Photo Credits

PLATES

1a, 1b, 2b, 3c, 3d, 4a–c, 5a–f, 7a–d, 8a–d, 9a–c, 10, 11a, 11c, 12a, 14a–d, 15a, 15c, 16a–c, 18a–c, 19, 20a, 21a–e, 22a–c, 23a, 23b, 24a, 24d, 25, 26a, 26b, 27d, 28, 29a–c, 31, 32, 33b, 33d, 34a–d, 35a, 35b, 35e, 35f, 36a, 36b, 37b, 38a, 38d–f, 39c, 40b, 41a, 42a, 42b, 43a–d, 44a, 44b, 45b, 45d, 46a, 46b, 49b–e, 51a–d, 52, 53a, 54a–c, 55a, 55b, 57a, 57b, 60a, 60b, 61a–61c, 63, 64, 65, 66, 68b, 68c, 70a, 70b, 72a, 73a, 73b, 74, 75a–d, 76b, 77, 79a, 79b, 80a, 80b, 81b, 82b, 82a, 83b, 84a, 84b, 85b, 86a, 86b, 87b, 88b, 89, 90a, 90b Ralph T. Hutchings   2a, 13f, 23c, 35c, 35g, 38b, 78a Patrick M. Timmons/Michael J. Timmons   3a, 17, 27a–c, 30, 33c, 36, 37a, 39a, 39d, 40a, 68a, 69a, 76a, 81a, 82a, 87a Mentor Networks, Inc.   6 Image provided by The Digital Cadaver™ Project, courtesy of Visible Productions, Inc.   11b, 12b–d, 13b–e, 20b, 56a–c, 58a–c, 72b, 78b–g, 86c, 87c   Frederic H. Martini, Inc.   13a, Pat Lynch/Photo Researchers, Inc.   13g, 24c, 33a, 39b, 54d, 59, 69b, 85a Custom Medical Stock Photo, Inc.   15b Martin M. Rotker   35d Thiem Verlagsgruppe   38c, 47a–d, 49a, 62a, 62b Pearson Education, PH College   42c, 44c   Wellcome Trust Medical   42d, 67 CNRI/Science Photo Library/Photo Researchers, Inc. Researchers, Inc.   45a David York/Medichrome/The Stock Shop, Inc.   48b Marconi Medical Systems, Inc.   48c, 50a, 71b, 78i www.NetAnatomy.com   50b ISM/Phototake NYC   50c, 53d Christopher J. Bodin, M. D., Tulane University Medical Center   53c Barry Slaven/P. Mode Photography/Photo Researchers, Inc.   53e P.M. Motta, A. Caggiati, G. Macchiarelli/Science Photo Library/Photo Researchers, Inc.   45c Science Photo Library/Photo Researchers, Inc. Researchers, Inc.   56d–f, 58d–f Visible Human Project/National Institutes of Health, National Library of Medicine   71a L. Basset/Visuals Unlimited   78h Michael L. Richardson; University of Washington of School of Medicine, Dept. of Radiology; www.rad.washington.edu